高校专门用途英语（ESP）系列教材

EST Reading
科技英语阅读教程

主编 陈 勇 廖华英
编者 官芬芬 鄢菁萍 廖莉莉 胡步芬

清华大学出版社
北京

内 容 简 介

本教材专门为科技英语课程编写。全书共12个单元,涉及核能与核辐射、机器人与人工智能、教育研究、地质与地球科学、生物技术、行为科学、材料学、工程学、计算机科学、信息技术、网络安全、科学与社会等较为热门的专业领域。每单元由课堂精讲的课文A和扩展阅读的课文B、C组成。每单元之前配有导读;课文A后面配有阅读理解、摘要写作、科技词汇、讨论等练习;课文B和C也编写了阅读理解和翻译等练习。此外,在教材末尾还附有常见数字、数字符号和数学式表达等专题知识和部分习题的参考答案。本书另配有PPT课件,需要的读者请访问www.tsinghuaelt.com下载使用。

本教材适用于科技英语专业学生、英语专业高年级学生,特别适用于理工类高校本科生和研究生,可作为科技英语、学术英语课程的必修课或选修课教材。

版权所有,侵权必究。举报:010-62782989,beiqinquan@tup.tsinghua.edu.cn。

图书在版编目(CIP)数据

科技英语阅读教程/陈勇,廖华英主编-北京:清华大学出版社,2017(2023.8重印)
(高校专门用途英语(ESP)系列教材)
ISBN 978-7-302-47917-8

Ⅰ.①科⋯　Ⅱ.①陈⋯　②廖⋯　Ⅲ.①科学技术-英语-阅读教学-高等学校-教材　Ⅳ.①N43

中国版本图书馆CIP数据核字(2017)第196397号

责任编辑:刘　艳
封面设计:平　原
责任校对:王凤芝
责任印制:刘海龙

出版发行:清华大学出版社
网　　址:http://www.tup.com.cn,http://www.wqbook.com
地　　址:北京清华大学学研大厦A座　　邮　编:100084
社 总 机:010-83470000　　邮　购:010-62786544
投稿与读者服务:010-62776969,c-service@tup.tsinghua.edu.cn
质量反馈:010-62772015,zhiliang@tup.tsinghua.edu.cn
印 装 者:大厂回族自治县彩虹印刷有限公司
经　　销:全国新华书店
开　　本:170mm×230mm　　印　张:22.25　　字　数:396千字
版　　次:2017年8月第1版　　印　次:2023年8月第7次印刷
定　　价:68.00元

产品编号:074055-03

前言
Foreword

在科学与技术飞速发展的今天，英语已成为全球范围内科技人员进行科学研究和学术交流必不可少的工具。查询与阅读科技文献、了解科技前沿发展状况、参加国内外学术交流，都离不开英语这一国际性语言。随着中国科技实力的进一步增长，中国人在科学与技术领域发出声音的机会越来越多，科技英语的重要性愈益凸显。作为英语的一种变体，科技英语有别于普通英语。科技英语在词汇、句子、语言风格等方面都有自身的特点，只有通过系统学习才能掌握。

当前，学生在完成基础阶段的大学英语课程之后，已经掌握了一定的语言知识和技能，但要应用于英语科技文献的阅读还有相当大的困难。为了填补普通英语与专业英语之间的空白，很多理工类高校开设了科技英语课程，试图在两者之间架设一座桥梁，让学生顺利进入专业英语的学习中。然而，由于科技英语不是主干课程，更由于大多数院校缺乏专门的科技英语教师，科技英语教学，尤其是教材建设面临很大的挑战。笔者在长期的科技英语教学实践中发现，在为数不多的已有教材中，科技英语的语言特征没有被有效地贯穿到学生的语言学习和训练中，还有改进的空间。正是在这样的背景下，我们组织了一批长期承担科技英语教学任务的教师编写了这本《科技英语阅读教程》。

本教材共12单元，每单元由同一学科领域的主课文A和扩展阅读课文B和C组成。课文主要节选自科技期刊 *Scientific American*、*Science*、*Nature* 等近期刊载的文章，其专业程度不超过非专业人士的理解能力，内容涉及核能与核辐射、机器人与人工智能、教育研究、地质与地球科学、生物技术、行为科学、材料学、工程学、计算机科学、信息技术、网络安全、科学与社会等较为热门的学科领域。教材末尾还附有常见数字、数字符号和数学式表达等专题知识。本教材编写的宗旨是满足理工类院校高年级本科生和硕士研究生学习科技英语的需要，重点培养其阅读理解的能力。

本教材具有以下几个独特的地方：

1. 科技英语词汇特点贯穿于整个练习编写中。练习的编写突出科技英语中科技术语、半科技术语的学习和掌握，促使学生充分认识一词多义现象，了解普通英语和科技英语的差别，并通过练习来掌握上述词汇特点。

2. 教材编写遵循学生自主学习的原则，课文后面没有编写词汇表，重点词汇全部融入词汇练习中，目的是让学生在教师的引导下自主查找并编写个性化词汇表，以解决学生过度依赖已有词汇表的通病，通过自主学习，养成良好的学习习惯。

3. 阅读材料量大是本教材的又一个特点。本书编者认为，任何一项严谨的科学研究都需要一个完整的过程，将这个过程呈现出来，不仅需要严密的逻辑思维，还需要充分的材料，所以，一定的篇幅是必要的。而学生通过阅读这些文章，不仅能掌握科技语言，更能学习科学的思维方式和研究方法。

4. 批判性思维在教材编写中的贯彻。本教材所选的阅读材料大多涉及不同甚至相反的观点，引导学生从多个角度看待同一个研究问题；与同类教材相比，本教材所选阅读材料除了来自通常意义上的理工科领域之外，还有来自人文社科领域的主题，甚至有科学与社会的关系等。这些材料的选择有利于训练学生的思辨能力。

5. 本教材提供了练习答案和PPT课件，作为教学参考。有需要课件的读者请访问ftp://ftp.tup.tsinghua.edu.cn/下载使用。

本教材适用于科技英语专业学生、英语专业高年级学生，特别适用于理工类高校本科生和研究生，可作为科技英语、学术英语课程的必修课或选修课教材，亦可作为英语爱好者学习和进修的参考书。

本书选取了近年来国外科技期刊的内容，我们首先要对原文作者表示最诚挚的谢意。本书的编写采用主编总体设计，编者各负其责的模式。在编写阶段，具体分工如下：陈勇负责第一单元至第五单元的编写；廖莉莉负责第六、十二单元；鄢菁萍负责第七、八单元；官芬芬负责第九单元至第十一单元；胡步芬负责附录的编写。策划阶段还得到了江西省质谱科学与仪器重点实验室、东华理工大学研究生院院长李满根教授的大力支持，清华大学出版社的编辑刘艳女士给予了悉心的指导和帮助，在此一并致以谢意。

作为一项有益的尝试，全体编写人员付出了辛苦与努力，但漏误在所难免，恳请使用本书的读者不吝赐教，以便将来再版时修正。

编者
2017年6月

目录
Contents

Unit 1 **Nuclear Power and Nuclear Radiation** 1
    Part I Intensive Reading 1
        Text A The Swallows of Fukushima......... 1
    Part II Extensive Reading 15
        Text B Laser Fusion, with a Difference 15
        Text C Five Years On from Fukushima 21

Unit 2 **Robots and Artificial Intelligence** 27
    Part I Intensive Reading 27
        Text A Robots with Heart 27
    Part II Extensive Reading 38
        Text B Launch the Nanobots 38
        Text C Should We Fear Supersmart Robots? 42

Unit 3 **Education Research** 46
    Part I Intensive Reading 46
        Text A A New Vision for Testing 46
    Part II Extensive Reading 61
        Text B The Science of Learning 61
        Text C Rebooting MOOC Research 69

Unit 4 **Geology and Geoscience** 75
    Part I Intensive Reading 75
        Text A The Oldest Rocks on Earth 75
    Part II Extensive Reading 89
        Text B Fossil GPS 89
        Text C The Birth of the Geological Map 96

Unit 5 **GM Technology** 100
    Part I Intensive Reading 100
        Text A Editing the Mushroom 100

	Part II	Extensive Reading	116
		Text B The American Chestnut's Genetic Rebirth	116
		Text C Saltwater Solution	124

Unit 6 Behavioral Science 130
Part I Intensive Reading 130
   Text A Good Habits, Bad Habits 130
Part II Extensive Reading 144
   Text B Baby Talk 144
   Text C Conquer Yourself, Conquer the World 152

Unit 7 Material Science 162
Part I Intensive Reading 162
   Text A Outshining Silicon 162
Part II Extensive Reading 176
   Text B Beyond Graphene 176
   Text C Using All Energy in a Battery 183

Unit 8 Engineering 187
Part I Intensive Reading 187
   Text A Birth of a Rocket 187
Part II Extensive Reading 201
   Text B Shape-Shifting Things to Come 201
   Text C Bottoms Up 209

Unit 9 Computer Science 218
Part I Intensive Reading 218
   Text A Extra Sensory Perception 218
Part II Extensive Reading 230
   Text B The Search for a New Machine 230
   Text C Just Add Memory 240

Unit 10 Information Technology 248
Part I Intensive Reading 248
   Text A Mind Games 248

	Part II	Extensive Reading	261
		Text B The Case of Stolen Words	261
		Text C The Tech Jungle	268

Unit 11 Cyber Security ... 271

Part I Intensive Reading ... 271
    Text A Saving Big Data from Itself ... 271

Part II Extensive Reading ... 282
    Text B How to Survive Cyberwar ... 282
    Text C In Tech We Don't Trust ... 288

Unit 12 Science and Society ... 291

Part I Intensive Reading ... 291
    Text A Our Transparent Future ... 291

Part II Extensive Reading ... 304
    Text B Privacy and Human Behavior in the Age of Information ... 304
    Text C What the "Right to Be Forgotten" Means for Privacy in a Digital Age ... 315

Appendix 常见数字、数字符号和数学式英语表达 ... **322**

Answer Keys 参考答案 ... **326**

Unit 1
Nuclear Power and Nuclear Radiation

导读

本单元主要涉及核能与核辐射研究。Text A以福岛核泄漏事件为样本，详细分析了低剂量核辐射对家燕的影响，并驳斥了部分研究者认为低剂量核辐射对环境生物无负面影响的观点，明确阐述了低剂量辐射对环境生物影响研究的意义。Text B则阐述了各种异于传统的聚变能研究方法和技术路线。Text C再次回到福岛核泄漏事件，从日本国内研究人员的视角来深度分析该事件暴露出来的日本科学界存在的严重问题，并提出解决办法。

Part I Intensive Reading

Text A

The Swallows of Fukushima

We know surprisingly little about what low-dose radiation does to organisms and ecosystems. Four years after the disaster in Fukushima, scientists are beginning to get some answers.

By Steven Featherstone

Until a reactor at the Chernobyl nuclear power plant exploded on April 26, 1986, spreading the equivalent of 400 Hiroshima bombs of fallout across the entire Northern Hemisphere, scientists knew next to nothing about the effects of radiation on vegetation and wild animals. The catastrophe created a living laboratory, particularly in the 1,100 square miles around the site, known as the exclusion zone.

In 1994 Ronald Chesser and Robert Baker, both professors of biology at Texas Tech University, were among the first American scientists allowed full access to the zone. "We caught a bunch of voles, and they looked as healthy as weeds. We became fascinated with that." Baker recalls. When Baker and Chesser sequenced the voles' DNA, they did not find abnormal mutation rates. They also noticed wolves, lynx and other once rare species roaming around the zone as if it were an atomic wildlife refuge. The Chernobyl Forum, founded in 2003 by a group of U.N. agencies, issued a report on the disaster's 20th anniversary that confirmed this view, stating that "environmental conditions have had a positive impact on the biota" in the zone, transforming it into "a unique sanctuary for biodiversity".

Five years after Baker and Chesser combed the zone for voles, Timothy A. Mousseau visited Chernobyl to count birds and found contradicting evidence. Mousseau, a professor of biology at the University of South Carolina, and his collaborator Anders Pape Møller, now research director at the Laboratory of Ecology, Systematics and Evolution at Paris-Sud University, looked in particular at Hirundo rustica, the common barn swallow. They found far fewer barn swallows in the zone, and those that remained suffered from reduced life spans, diminished fertility (in males), smaller brains, tumors, partial albinism—a genetic mutation—and a higher incidence of cataracts. In more than 60 papers published over the past 13 years, Mousseau and Møller have shown that exposure to low-level radiation has had a negative impact on the zone's entire biosphere, from microbes to mammals, from bugs to birds.

Mousseau and Møller have their critics, including Baker, who argued in a 2006 *American Scientist* article co-authored with Chesser that the zone "has effectively become a preserve" and that Mousseau and Møller's "incredible conclusions were supported only by circumstantial evidence".

Almost everything we know about the health effects of ionizing radiation comes from an ongoing study of atomic bomb survivors known as the Life Span Study, or LSS. Safety standards for radiation exposures are based on the LSS. Yet the LSS leaves big questions about the effects of low-dose radiation exposure unanswered. Most scientists agree that there is no such thing as a "safe" dose of radiation, no matter how small. And the small doses are the ones we understand the least. The LSS does not tell us much about doses below 100 millisieverts (mSv). For instance, how much radiation does it take to cause genetic mutations, and are these mutations heritable? What are the mechanisms and genetic biomarkers for radiation-induced diseases such as cancer?

The triple meltdown at the Fukushima Daiichi nuclear power plant in March 2011 created another living lab where Mousseau and Møller could study low doses of radiation, replicating their Chernobyl research and allowing them "much higher confidence that the impacts we're seeing are related to radiation and not some other factor," Mousseau says. Fukushima's 310-square-mile exclusion zone is smaller than Chernobyl's but identical in other ways. Both zones contain abandoned farmland, forests and urban areas where radiation levels vary by orders of magnitude over short distances. And they would almost certainly gain access to Fukushima more quickly than scientists could get into Soviet-run Chernobyl. In short, Fukushima presented an opportunity to settle a debate.

Within months of Fukushima, Mousseau and Møller were counting birds in the contaminated mountain forests west of the smoldering nuclear plant, but they could not get into the zone itself to see what was happening to the barn swallows. Finally, in June 2013, Mousseau was among the first scientists allowed full access to Fukushima's exclusion zone.

Sensitivity to radiation varies greatly in living things and among individuals of the same species, which is one reason it is important not to extrapolate from butterflies to barn swallows or from voles to humans. Butterflies are particularly radiosensitive, Mousseau says. In August 2012 the online journal *Scientific Reports* published a paper examining the effects of Fukushima's fallout on the pale grass blue butterfly. Joji Otaki, a biology professor at the University of the Ryukyus in Okinawa, revealed that butterflies collected near Fukushima two months after the disaster had malformed wings, legs and eyes. Mousseau and Møller's surveys of insects in Chernobyl and Fukushima show drop-offs in butterflies as a group. But Otaki's paper adds an important new wrinkle. When he bred mutant Fukushima butterflies with healthy lab specimens, the rate of genetic abnormalities increased with each new generation.

Mousseau believes that this phenomenon, the accumulation of genetic mutations, is a hidden undercurrent eroding the health of radioactive ecosystems, occasionally revealing itself in the offspring of mutant butterflies or barn swallows with partial albinism. Even Baker agrees with Mousseau on Otaki's conclusions: "Clearly, there's something going on with the butterflies that's radiation-induced. Multigenerational exposure does result in an altered genome."

I met Mousseau and his postdoctoral fellow, an Italian named Andrea Bonisoli Alquati, at the airport and then we drove to our hotel in Minamisoma, north of the Fukushima power plant. We passed through one deserted town after the next, meandering north toward the nuclear plant. Mousseau scanned shuttered storefronts and empty houses for barn swallow nests as he drove. Barn swallows are ideal scientific subjects because they are philopatric, meaning the birds tend to return to breed in the same locations over a lifetime. Much is already known about them under normal conditions, and they share similar genetic, developmental and physiological characteristics with other warm-blooded vertebrates. The barn swallow is the proverbial canary in the coal mine, except the coal mine in question is radioactive❶. Mousseau counted about a dozen old nest "scars", crescent-shaped blots of mud plastered under eaves, but not one new nest.

"They were showing such negative effects the first year," he said. "I figured it'd be very difficult to find them this year."

"I just can't believe there aren't any active barn swallow nests. I don't see any butterflies flying. Don't see any dragonflies flying. It's really a dead zone." he said.

Fukushima offers a vanishingly rare glimpse of an ecosystem's early response to radioactive contamination. Little is known about generations of Chernobyl's voles and barn swallows, not to mention other critters. Anecdotal reports point to massive die-offs of plants and animals, but no details exist about their recovery. Did some species evolve a heightened ability to repair DNA damaged by radiation? Studying Fukushima's ecosystem, right now, is critical to developing predictive models that could explain how adaptations to low-level radiation exposure, as well as the

---

❶ The barn swallow is the proverbial canary in the coal mine, except the coal mine in question is radioactive. 金丝雀对煤矿瓦斯特别敏感。当瓦斯含量超过一定限度而人还未觉察时，金丝雀已经毒发身亡。这里作者借用了金丝雀来打比方，即家燕对于核辐射的敏感如同煤矿里金丝雀之于瓦斯的敏感。

accumulation of genetic damage, progress over time.

Mousseau regretted that he could not get access to the zone immediately after the accident. "We'd have much more rigorous data on how many swallows were there, how many disappeared," he said after we arrived at the hotel. "Are the ones that are coming back the resistant genotypes, or are they just lucky in some way?"

The next day, with Mousseau's permits validated, a line of officers waved our car through the barricades and into the exclusion zone. Mousseau planned to work his way along the coastal plain, counting every barn swallow, plotting the location of every nest and capturing as many of the birds as possible. "Every data point we get here is absolutely invaluable," he said to Bonisoli Alquati.

A mile from the nuclear plant Bonisoli Alquati spotted a barn swallow perched on a wire near a house. A nest made with fresh mud sat on a ledge inside the garage. Radiation levels peaked at 330 microsieverts per hour, more than 3,000 times above normal background radiation and the highest level Mousseau has ever recorded in the field. "In 10 hours, you'll get your annual dose," said Bonisoli Alquati, referring to the amount of background radiation the average person in the U.S. receives in an entire year.

Futaba is a ghost town, off-limits to all except former residents, who are allowed to return for only a few hours every month to check on homes and businesses. A sign over the town's commercial center reads, "Nuclear Power: Bright Future of Energy." Radiation levels on the main street were no worse than many contaminated areas outside the zone. Peering through binoculars, Kitamura counted six swallows circling near a smashed sporting goods shop.

"Set up the nets and poles!" he shouted.

Kitamura and Bonisoli Alquati crouched outside the store, a mist net bunched loosely between them. Swallows swooped and chattered overhead. Bird by bird, it took two hours to catch and sample all six swallows. Before releasing the birds, Mousseau fitted them with tiny thermoluminescent dosimeters (TLDs) to track their radiation dose. Down by the Futaba train station, where radiation levels were 10 times higher, they captured two more swallows.

The Japanese government initially vowed to clean up 11 of the most severely contaminated municipalities in Fukushima Prefecture by March 2014. Their goal was to reduce annual dose rates to 1 mSv, the limit for the general public, according to the recommendations of the International Commission on Radiological Protection. But the bulk of the cleanup effort has so far been focused on stabilizing the damaged reactors at the nuclear plant, which continue to leak radiation into the Pacific. Japanese authorities no longer have a specific time frame for decontamination. Instead they have set 1 mSv per year as a long-term goal and are now encouraging some of the 83,000 evacuees to return to places with annual dose rates of up to 20 mSv, equivalent to the commission's dose limit for nuclear workers. The ruling party in Japan recently issued a report acknowledging that many contaminated areas will not be habitable for at least a generation.

This goalpost moving underscores the gap between our knowledge of the effects of low-dose radiation and public policy governing—among other things—nuclear cleanup protocols. Although scientists have not determined a "safe" dose of radiation, Japanese administrators need a target number to craft decontamination and resettlement policies, so they rely on advisory bodies such as the International Commission on Radiological Protection and imperfect studies such as the LSS.

Brenner's research shows evidence for increased rates of cancer associated with annual doses as low as 5 mSv. Below this arbitrary threshold, there is no firm evidence for or against direct health risks in humans, although Mousseau and Møller have observed negative effects in plant and animal populations. "Once you get down to these sorts of doses, you have to rely on best understandings of mechanisms," Brenner says, "and that's pretty limited."

In a residential neighborhood on the outskirts of Namie, Bonisoli Alquati spotted a barn swallow nest wedged in a narrow alley between two houses. It was the first active nest he had seen after a disappointing day of cruising the deserted districts around Futaba and Namie, counting dozens of empty nests and scars. Counting nests before the rain washes them all away is crucial to establishing a baseline for what swallow populations were before the accident, but Mousseau also needed samples from live birds for his lab work. The nest in the alley contained three chicks, the first he found in the zone, and three undeveloped eggs. "This is an important nest,"

Unit 1
Nuclear Power and Nuclear Radiation

Mousseau said.

Bonisoli Alquati sat in the front seat of the car. He scooped a chick out of a plastic container and measured it with various tools. Puffing on the downy underside of the chick's wing, he exposed a patch of skin and lanced it with a needle. Some of the blood went into a capillary tube; some got smeared on a glass slide. Then he cinched the chick in a canvas sack and lowered it into the "oven", a stack of lead bricks strapped together with duct tape. The bricks formed a shielded chamber, allowing Mousseau to measure the whole-body burden of individual birds without background radiation muddying the result.

"Our objective is to be able to look at individual birds from one year to the next and to determine whether the probability of survival is related to the dose they receive," he said. "If we really want to get at mechanisms of genetic variation and radio-sensitivity and how they impact individuals, then it's necessary to do this finer-scale dosimetry."

But radiation levels in this spot were too hot for accurate measurements. Mousseau moved the car down the street and reset the gamma spectrometer. After a few minutes, it displayed a distinct signal for cesium 137 contamination, the main isotope in Fukushima's fallout. The chick, perhaps a week old, was radioactive.

Barn swallows are omens of good fortune in Japan. Many people nail little wooden platforms over the doors of their houses to attract the birds. In the zone, the platforms, like the houses, were all empty. Each day after the zone closed, Mousseau and Bonisoli Alquati worked well into the night, capturing barn swallows in clean areas north of Fukushima to establish a control group. Clean is a relative term. Background radiation in Minamisoma, which was evacuated during the disaster, is still twice that of normal. It was strange to find barn swallow nests overflowing with fat, peeping chicks.

On Mousseau's last day in Japan, he spotted an active barn swallow nest on a gritty side street in Kashima. Mousseau received permission from a neighbor to net the birds. A member of the local river society, he said he was glad somebody was investigating the radioactive contamination because the government was not. "Always secret, the government," he said, complaining about fallout washing into the river.

Koi fish caught there registered 240,000 becquerels of cesium per kilogram, he said. People do not eat these fish, which is fortunate, because the radiation limit for fish consumption in Japan is 100 becquerels per kilogram.

Forty percent of us will one day be diagnosed with some form of cancer. If there is a signal hidden in the noise of this sobering statistic, one that might point to low-dose radiation-induced cancers, it is too faint for epidemiologists to hear. The big questions about low-dose radiation will eventually be answered by researchers studying "radiation-induced chromosome damage, or radiation-induced gene expression, or genomic instability," Brenner says. This is the direction Mousseau and Møller are beginning to take with their research on barn swallows.

"Unfortunately, tumors don't tell us if they were caused by radiation or something else," Mousseau says. If he had enough funding, Mousseau would sequence the DNA of every swallow that he fitted with a TLD in the field. By comparing the results with individual dose estimates, he might be able to locate genetic biomarkers for radiation-induced diseases.

Last November, Mousseau made his 12th trip to Fukushima, 18 months after I accompanied him to the zone. Mousseau and Møller have published three papers demonstrating steep declines in Fukushima's bird populations. Mousseau says that the latest census data, which they are preparing to publish in the *Journal of Ornithology*, provide "pretty striking" evidence for continued declines, "with no evidence of a threshold effect." But for some reason, radiation appears to be killing off birds in Fukushima at twice the rate it is in Chernobyl. "Perhaps there is a lack of resistance, or there is an increased radiosensitivity in Fukushima's native populations," Mousseau says. "Perhaps Chernobyl birds have evolved resistance to some degree, or the ones that are susceptible have been weeded out over the past 26 years. We don't really know the answer to that, but we're hoping to get to it." The answer might be found in the blood of the barn swallows that Mousseau and Bonisoli Alquati collected on our trip. A preliminary analysis of those samples does not reveal any evidence for a significant increase in genetic damage, although it is still too early to tell. Mousseau needs many more samples from barn swallows in the most contaminated areas, where populations are crashing.

Although Mousseau and Møller's initial findings afford a compelling glimpse of a troubled ecosystem in Fukushima, the 2014 report by the U.N. Scientific Committee on the Effects of Atomic Radiation (UNSCEAR) echoes its earlier assessment of the Chernobyl disaster, declaring that radiation effects on "nonhuman biota" in highly contaminated areas are "unclear" and are "insignificant" in less contaminated ones.

"We're doing basic science, not toxicology, but UNSCEAR hasn't gone to the trouble of either asking us about our work or finding someone to interpret our findings," Mousseau says. "They set the standard for human health, and they're ignoring a large portion of potentially relevant information."

He says the evidence being ignored is substantial. "In my years of experience at Chernobyl and now Fukushima, we've found signals of the effects of increased mutation rates in almost every species and every network of ecological processing that we've looked at," Mousseau says.

Baker has no plans to conduct research in Fukushima, but he recently sequenced DNA from a different genus of vole from Chernobyl. The new data appear to support Mousseau's and Otaki's conclusions that elevated mutation rates are linked to radiation exposure. The consequences of multigenerational exposure, whether or not it diminishes an animal's fitness or reproductive capabilities or causes birth defects or cancers in future generations, are still unclear.

(Excerpt from *Scientific American*, February 2015)

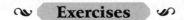

I. Reading Comprehension

● Section One

Directions: Answer the following questions based on the information from the text.

1. What was Ronald Chesser and Robert Baker's opinion about radioactive impact on the biota? What was the contradicting evidence Timothy A. Mousseau found in Chernobyl five years later? Compare their interpretations and draw your own conclusions.

2. What are the similarities between the exclusion zones of Chernobyl and Fukushima?
3. Why did Mousseau study barn swallows?
4. What is the problem of Life Span Study (LSS)?
5. What is the significance of studying Fukushima's ecosystem?

● **Section Two**

Directions: Write an abstract based on the text in no more than 200 words.

Abstract:
Key words:

II. Vocabulary

● **Section One**

Directions: Choose the explanation that is closest in meaning to the underlined part in each sentence.

1. The Chernobyl Forum issued a report on the disaster's 20th anniversary that confirmed this view, stating that "environmental conditions have had a positive impact on the biota" in the zone, transforming it into "a unique sanctuary for biodiversity".

   A. a nature reserve B. immunity from arrest
   C. a holy place   D. a place for unwanted animals of a specified kind

2. Five years after Baker and Chesser combed the zone for voles, Timothy A. Mousseau visited Chernobyl to count birds and found contradicting evidence.

   A. wrong   B. opposing   C. inconsistent   D. denying

Unit 1
Nuclear Power and Nuclear Radiation

3. Fukushima's 310-square-mile exclusion zone is smaller than Chernobyl's but <u>identical</u> in other ways.
   A. derived from a single fertilized egg or ovum
   B. coinciding exactly when superimposed
   C. similar in every detail; exactly alike
   D. having properties with uniform values along all axes

4. Sensitivity to radiation varies greatly in living things and among individuals of the same species, which is one reason it is important not to <u>extrapolate</u> from butterflies to barn swallows or from voles to humans.
   A. draw from specific cases for more general cases
   B. estimate the value of
   C. gain knowledge
   D. infer unknown data from known data

5. Little is known about generations of Chernobyl's voles and barn swallows, not to mention other <u>critters</u>.
   A. living creatures   B. domestic animals   C. persons   D. human beings

6. The barn swallow is the <u>proverbial</u> canary in the coal mine, except the coal mine in question is radioactive.
   A. relating to a proverb   B. widely known and spoken of
   C. confessed   D. notorious

7. We'd have much more <u>rigorous</u> data on how many swallows were there, how many disappeared.
   A. severe   B. tight
   C. thorough and strict   D. careful and thorough

8. If there is a signal hidden in the noise of this <u>sobering</u> statistic, one that might point to low-dose radiation-induced cancers, it is too faint for epidemiologists to hear.
   A. cause to become more serious and sensible
   B. not affected by alcohol
   C. completely lacking in playfulness
   D. dull; lacking in color

9. Although Mousseau and Møller's initial findings afford a compelling glimpse of a troubled ecosystem in Fukushima, the 2014 report by the U.N. Scientific Committee on the Effects of Atomic Radiation echoes its earlier assessment of the Chernobyl disaster.
   A. imitates   B. continues to be discussed and remained influential
   C. rings with sound   D. repeats and expresses agreement with others' opinion
10. He says the evidence being ignored is substantial.
    A. fairly large   B. stout
    C. of real importance or validity   D. solidly built

● **Section Two**

Directions: There are two or three meanings for each semi-technical term underlined in the following sentences. Choose the correct one according to the context.

1. Until a reactor at the Chernobyl nuclear power plant exploded on April 26, 1986, spreading the equivalent of 400 Hiroshima bombs of fallout (radiation that affects a particular area after a nuclear explosion has taken place; unpleasant consequences of something that has happened) across the entire Northern Hemisphere, scientists knew next to nothing about the effects of radiation on vegetation and wild animals.

2. When Baker and Chesser sequenced the voles' DNA, they did not find abnormal mutation (change or alteration in form or qualities; alteration in the inherited nucleic acid sequence of the genotype of an organism) rates.

3. Those that remained suffered from reduced life spans, diminished fertility (the ratio of live births in an area to the population of that area; the capability of producing offspring; the property of producing abundantly and sustaining vigorous growth) in males, smaller brains, tumors, partial albinism, and a high incidence (the frequency with which something occurs; the striking of a light beam on a surface) of cataracts.

4. Yet the LSS leaves big questions about the effects of low-dose radiation exposure (publicity a person receives; harmful effect on one's body caused by very cold weather; being in a situation where something dangerous might affect you) unanswered.

5. The triple meltdown (the sudden and complete failure of a company, organization, or system; severe overheating of the core of a nuclear reactor resulting in the core melting and radiation escaping) at the Fukushima Daiichi nuclear power plant in March 2011

Unit 1
Nuclear Power and Nuclear Radiation

created another living lab.

6. The triple meltdown at the Fukushima Daiichi nuclear plant created another living lab where Mousseau and Møller could study low doses of radiation, replicating (reproducing a copy of genetic material or a living organism itself; repeating a scientific experiment to obtain a consistent result; bending or turning backward) their Chernobyl research and allowing them "much higher confidence that the impacts we're seeing are related to radiation and not some other factor", Mousseau says.

7. Counting nests before the rain washes them all away is crucial to establishing a baseline (the line marking each end of a tennis or badminton or volleyball court; a minimum or starting point used for comparison) for what swallow populations were before the accident.

8. Are the ones that are coming back the resistant (relating to or conferring immunity to disease or infection; engaged in defiance of established authority) genotypes, or are they just lucky in some way?

9. He planned to work his way along the coastal plain, counting every barn swallow, plotting (planning secretly to do something that is illegal or wrong, usually against a person or government; carefully planning each step of a strategy or a course of action; marking a position on a map using instruments to obtain accurate information) the location of every nest and capturing as many of the birds as possible.

10. Koi fish caught there registered (indicated a reading of gauges and instrument; recorded in writing or entered into a book of names; showed in one's face) 240,000 becquerels of cesium per kilogram.

● **Section Three**

Directions: Match the Chinese terms with their English equivalents.

1. 基因突变   A. exclusion zone
2. 辐射诱导的   B. biota
3. 对辐射敏感的   C. biodiversity
4. 禁区   D. circumstantial evidence
5. 放射性剂量计   E. genetic mutation
6. 生物区   F. ionizing radiation

7. 预测模型　　　　　　　G. radiation-induced
8. 生物多样性　　　　　　H. genetic biomarker
9. 数量级　　　　　　　　I. order of magnitude
10. 间接证据　　　　　　　J. radiosensitive
11. 实验对照组　　　　　　K. dosimeter
12. 基因生物标志　　　　　L. predictive model
13. 生理学特征　　　　　　M. gamma spectrometer
14. 伽马光谱仪　　　　　　N. control group
15. 电离辐射　　　　　　　O. physiological characteristic

III. Questions for Discussion

Directions: Work in groups and discuss the following questions.

1. What is your attitude towards nuclear power? Are you for or against it? Support your opinion with sound facts and figures.
2. Should regulators make it public if a nuclear accident happened? Why?
3. Which factor caused the meltdown at the Fukushima Daiichi nuclear power plant in 2011, the management or the technological failure?

Unit 1
Nuclear Power and Nuclear Radiation

Part II Extensive Reading

Text B

Laser Fusion, with a Difference

Europe's Laser Mégajoule project blazes its own trail toward nuclear ignition.

By Daniel Clery

To anyone familiar with laser fusion research, a visit to Laser Mégajoule (LMJ❶), a €3 billion research facility completed late last year near France's Atlantic coast, triggers instant déjà vu. The site is a dead ringer for the world's leading lab, the National Ignition Facility (NIF❷) in California. LMJ has the same stadium-sized building, the same shiny white metal framework, the same square beam tubes and 10-meter-wide reaction chamber. The security is less obtrusive, the visitors' area larger and more informative. Overall, however, walking through LMJ's doors is like stepping into a parallel universe in which Lawrence Livermore National Laboratory, the U.S. nuclear weapons laboratory that runs NIF, has somehow come under French control.

The similarities are no coincidence. Both sites were designed for the same purpose—to train scores of powerful laser beams on a single target, subjecting it, for an instant, to outlandish extremes of temperature and pressure. The two labs have collaborated extensively, and the primary mission of each is military: replicating nuclear explosions in miniature so that weapons scientists can ensure their bombs will detonate if needed without having to test them. The French facility, like its U.S. counterpart, will also pursue a sideline in inertial fusion energy (IFE) research: crushing capsules of hydrogen isotopes with laser pulses so that the isotopes fuse into helium, releasing vast stores of energy that might one day be harnessed in a power plant.

---

❶ LMJ: Laser Mégajoule (法)兆焦耳激光器
❷ NIF: National Ignition Facility (美)国家点火装置

But in a major departure from NIF's initial approach, LMJ is putting top-secret weapons research first. Once NIF was complete in 2009, Livermore researchers immediately embarked on a crash program to achieve ignition—generating a self-sustaining fusion reaction that produces as much energy as went into triggering it. They failed to reach that goal and have since changed their approach.

France's Alternative Energy and Atomic Energy Commission (CEA❶), which built LMJ, also wants to achieve ignition, because it's key to both weapons research and energy. "The target of ignition drives the design of the machine," says Pierre Vivini, LMJ's project leader. But generating power from laser fusion will be left to outside academic researchers, and they won't get their hands on the machine for another 2 years.

At their core, the two facilities are near-twins. Just as at NIF, LMJ researchers use a fiber laser to produce a pulse of infrared light that lasts a few billionths of a second, with just billionths of a joule of energy. This weak pulse then passes into preamplifiers, slabs of neodymium-doped glass that are pumped full of energy by xenon flash lamps just before the pulse comes through. They dump that energy into the beam, boosting it to about a joule, before the light is split into many parallel beams and sent to the main amplifiers (the same neodymium glass and flash lamps, only bigger).

LMJ has 22 main amplifier chains, arranged in four vast halls around the building, and each amplifier accommodates eight parallel beams at a time. During a laser shot, the eight beams are bounced back and forth through the amplifier four times to multiply their energy by a factor of 20,000. An elaborate array of mirrors will direct the 176 beams around all sides of the spherical reaction chamber; then a final set of optics will convert the beams from infrared to ultraviolet (UV) light and focus them to a needle-sharp point at the center of the chamber. Recombining all the beams delivers 1.5 megajoules of energy into the target at the center of the chamber—roughly the same as the kinetic energy of a 2-tonne truck traveling at 140 kilometers per hour. NIF's laser delivers 1.8 megajoules.

---

❶ CEA: France's Alternative Energy and Atomic Energy Commission 法国替代能源与原子能委员会

Unit 1
Nuclear Power and Nuclear Radiation

Because of funding constraints, only one main amplifier chain is now online. But that is enough to get started on nuclear weapons research, says François Geleznikoff, director of nuclear weapons at CEA: "With eight beams we can do good physics. We don't need all the beams to study weapons." The facility will add at least another two chains (16 beams) each year until it reaches full capacity sometime in the next 10 years. Well before then, academic IFE researchers from across Europe have been promised at least 20% of the machine's time.

Their game plan includes some significant deviations from NIF's strategy. For example, they've won funding for a separate laser to be installed at LMJ, providing pulses much shorter and more powerful than anything NIF can match. The PETawatt Aquitaine Laser(PETAL[1]) will generate pulses with a relatively modest 3.5 kilojoules of energy, but that energy will be crammed into a trillionth of a second, producing a power of more than a thousand trillion watts—a hundred times that of LMJ's pulses. PETAL's pulses won't be split and delivered from all directions; they'll come from a single direction timed to coincide with a pulse from the main laser. In experiments, they could provide a sudden intense kick of power or could be used like a strobe light to take snapshots of what's going on.

Experiments combining PETAL and LMJ will mimic the conditions in the interiors of stars and other astrophysical objects. Researchers will also use the powerful laser jolts to accelerate protons—an approach that could yield compact accelerators for cancer therapy. But what excites laser fusion researchers is the prospect that the short, sharp blasts from PETAL could act as a spark plug for fusion reactions.

The hope is that using PETAL in this way will allow IFE researchers to avoid some of the pitfalls that have hobbled NIF. The key element of any IFE scheme is the fuel capsule, a plastic sphere about the size of a peppercorn containing frozen deuterium and tritium—isotopes of hydrogen that are the fuel of fusion. Placed at the center of the reaction chamber, the plastic of the capsule is vaporized by the intense heating from the laser pulse, causing an implosion that crushes the fuel to 100 times

---

[1] PETAL: The PETawatt Aquitaine Laser 一项计划安装在LMJ装置上的单独的激光资助项目

the density of lead and heats it to 100 million K—which should be sufficient for fusion to ignite.

At NIF and in the weapons research experiments at LMJ, researchers trigger the implosion indirectly by enclosing the capsule in a metal can that is heated by the laser and in turn bombards the capsule with X-rays. That approach offers some advantages—it smooths out imperfections in the laser beam, and X-rays are better than UV light at driving the implosion—but it makes the target complex and expensive, not what you want for energy generation. NIF researchers have struggled to make this approach work: Energy is lost in the process of converting light into X-rays, and the implosions do not progress smoothly.

IFE researchers outside the weapons labs want to do things differently. By getting rid of the can and targeting beams directly on the capsule, they can avoid the complications and energy loss of converting UV light to X-rays. To get a smooth, symmetrical implosion, many advocate driving it more slowly. But then the compressed fuel won't get hot enough to start reacting on its own; it will need an extra spark to start it off.

One possible solution, known as fast ignition and pioneered at the Osaka University in Japan, is to use a short, high-powered laser pulse as the spark—the sort of pulse that PETAL can produce. Last decade, European IFE researchers proposed building a demonstration IFE reactor based on fast ignition, called HiPER. That plan lost some momentum when NIF fell far short of ignition, but its proponents hope LMJ-PETAL will give it new impetus.

Recent experiments on lower powered machines have suggested that PETAL might not pack enough punch to trigger fast ignition. But another alternative approach, pioneered at the University of Rochester in New York, might save the day. The technique, known as shock ignition, compresses the fuel capsule with a laser pulse from the main laser as in other techniques. But at the end of the pulse, the laser adds a sudden spike of power to produce a shock wave converging on the center of the fuel. When the shock hits the center, the sudden hike in pressure sparks the reaction. "Experiments at Omega [Rochester's laser] and elsewhere are encouraging, and the laser requirements [for shock ignition] look rather more benign than fast

Unit 1
Nuclear Power and Nuclear Radiation

ignition at this stage," says Chris Edwards, a fusion researcher at the Central Laser Facility at the United Kingdom's Rutherford Appleton Laboratory and one of the leaders of HiPER.

In their quest for fusion energy with LMJ-PETAL, researchers face sociological and political challenges, too. The European IFE community is small and is not used to working with such a huge machine or with weapons lab levels of security. "LMJ alone is like a cathedral in the desert," Batani says. "Researchers are interested, but suspicious. Many are not convinced it is a good tool for research." CEA also needs to overcome its reluctance to share simulation codes with academic researchers for fear of helping rogue nations develop thermonuclear weapons, Batani says: "We need reliable simulations, but there is no open code." And Europe has traditionally focused on a different approach, magnetic confinement fusion, which has its own cathedral not far away: the multibillion-euro ITER, under construction in Cadarache, France. "If shock ignition on LMJ works, politicians could become more positive," Batani says. And Europe—always an also-ran in this branch of fusion energy—might just gain some boasting rights.

(Excerpt from *Science*, January 9, 2015)

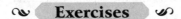

I. Translate the following technical terms into English.

1. 反应室 _____
2. 激光束 _____
3. 氢同位素 _____
4. 自持聚变反应 _____
5. 光纤激光器 _____
6. 前置放大器 _____
7. 动能 _____
8. 火花塞 _____
9. 模拟程序 _____
10. 内爆 _____
11. 热核武器 _____
12. 燃料芯块 _____
13. 脉冲氙灯 _____
14. 冲击点火 _____
15. 高能激光脉冲 _____
16. 紫外线 _____

17. 惯性聚变能 _____   18. 磁约束核聚变 _____
19. 电子闪光灯 _____   20. 氘/氚 _____

II. Translate the following paragraphs into Chinese.

Experiments combining PETAL and LMJ will mimic the conditions in the interiors of stars and other astrophysical objects. Researchers will also use the powerful laser jolts to accelerate protons—an approach that could yield compact accelerators for cancer therapy. But what excites laser fusion researchers is the prospect that the short, sharp blasts from PETAL could at as a spark plug for fusion reactions.

The hope is that using PETAL in this way will allow IFE researchers to avoid some of the pitfalls that have hobbled NIF. The key element of any IFE scheme is the fuel capsule, a plastic sphere about the size of a peppercorn containing frozen deuterium and tritium—isotopes of hydrogen that are the fuel of fusion. Placed at the center of the reaction chamber, the plastic of the capsule is vaporized by the intense heating from the laser pulse, causing an implosion that crushes the fuel to 100 times the density of lead and heats it to 100 million K—which should be sufficient for fusion to ignite.

At NIF and in the weapons research experiments at LMJ, researchers trigger the implosion indirectly by enclosing the capsule in a metal can that is heated by the laser and in turn bombards the capsule with X-rays. That approach offers some advantages—it smooths out imperfections in the laser beam, and X-rays are better than UV light at driving the implosion—but it makes the target complex and expensive, not what you want for energy generation. NIF researchers have struggled to make this approach work: Energy is lost in the process of converting light into X-rays, and the implosions do not progress smoothly.

Unit 1
Nuclear Power and Nuclear Radiation

Text C

Five Years On from Fukushima

To build sustainability and trust, energy and environment research in Japan must become more interdisciplinary and global, say Masahiro Sugiyama and his colleagues.

By Masahiro Sugiyama et al.

Next week will mark five years since 11 March 2011, the day of the devastating Tohoku earthquake and tsunami, and the accident that followed at the Fukushima Daiichi nuclear-power plant of the Tokyo Electric Power Company. The quake and tsunamis killed nearly 16,000 people and injured more than 6,000; 2,600 are still missing.

Fukushima prefecture, the location of the crippled nuclear plant, was hit particularly hard. The Japanese government provisionally rated the severity of the accident on par with the 1986 Chernobyl disaster—a seven on the seven-point International Nuclear and Radiological Event Scale. Around 110,000 people had to evacuate because of dispersed radionuclides. Despite the large-scale decontamination efforts, about 70,000 former residents are yet to return.

Shocked by the fallout, Japan changed its energy policy. The year before the disaster, the 2010 Basic Energy Plan had called for 53% of electricity generation to come from nuclear power by 2030, implying significant new construction. Since the accident, Japan's energy policy has featured an expanded role for renewables and market liberalization—transitioning from a regional-monopoly model to one that is open to competition.

In July 2012, reflecting the public's desire for a transition towards renewable energy, Japan introduced a feed-in tariff, which guarantees renewable-energy generators a high price for the electricity that they feed into the grid; it was particularly generous for solar photovoltaics. Installed solar capacity more than quadrupled in the first three years. The government also strengthened nuclear security substantially. A newly created, independent Nuclear Regulation Authority instituted new safety measures in 2013. All of Japan's nuclear-power plants halted operation for inspection after the accident, and the share of nuclear in power generation dropped from about 30% in 2010 to zero in 2014. Only 4 of the 44 reactors have been restarted so far.

The public debate on nuclear rumbles on. Still an important resource for this resource-poor nation, it is expected to provide 20%–22% of electricity by 2030, under a new long-term energy strategy created in conjunction with Japan's pledge to the United Nations to cut greenhouse-gas emissions.

- **Weak Links**

The journey since 2011 has been difficult, with policy controversies on every front. Many of the fraught decisions—on evacuation, clean-up, energy transition and disaster preparedness—were at the science-policy interface. And scientists, especially those involved in giving policy advice, lost credibility and the trust of the public.

Several initiatives have been launched to rebuild these crucial bridges. One such is an effort by the Japan Science and Technology Agency (JST) on research into scientific advice, and a "deliberative polling" exercise in August 2012 involving the public that was used to inform the energy policy of the previous government. These efforts are yielding genuine progress, but slowly.

We strongly believe that the events and aftermath of 11 March highlighted a fundamental problem with research in Japan: weak connections between disciplines and between Japan's scholars and those working in other countries. In a nation that performs world-class research in conventional disciplines, interdisciplinary scholarship lags, and Japanese researchers are keenly aware of this. Moreover, the nation's breadth of disciplinary coverage is narrower and the rate of international collaboration is lower than in comparable nations.

This has a particularly important implication for energy and environment research, which requires the integration of diverse knowledge that can come from anywhere in the world. During the Fukushima crisis, researchers who were not used to collaborating with other disciplines (or other nations) struggled to do so.

Unit 1
Nuclear Power and Nuclear Radiation

● **Two Case Studies**

Two examples illustrate the problem. The first concerns assessing the safety of nuclear power plants. Probabilistic risk assessment (PRA) is a standard tool to quantitatively evaluate the likelihood of severe accidents and their impacts, using analysis methods such as fault trees and Monte Carlo simulation. Before the earthquake, nuclear experts conducting PRA research in Japan focused on internal events at nuclear plants (mechanical component failures and human errors), dealing mostly with engineering knowledge.

What the disaster vividly demonstrated is that nuclear plants are susceptible to external events and that accident impacts are not contained—they may include the release of radionuclides, with dire environmental effects. The PRA has therefore been extended beyond nuclear engineering to cover disciplines ranging from seismology and geology to atmospheric science and ecological modelling.

Before 2011, such interdisciplinary PRA research in Japan was limited compared with other developed economies that have significant nuclear presence such as in the United States, the United Kingdom and France. This was partly because the country did not require PRA for regulatory purposes. Under the new 2013 regulations, Japan mandated the use of PRA for nuclear plants and is now trying to catch up in this area.

The second example concerns innovation in renewable energy. Although many citizens would prefer the nation's energy portfolio to have a larger share of renewables, these are more costly in Japan than elsewhere, even for technologies such as solar photovoltaics, in which Japan was a pioneer.

Ideally, Japan should explore how each policy alternative might affect future cost trends for solar. Combining energy-systems analysis and policy analysis with technology forecasting methods would yield crucial insights. Such interdisciplinary studies in Japan are hard to come by. Because of the low level of global networking in these areas, international experiences are not widely appreciated in Japan either.

This has already affected the solar market in Japan. Under the feed-in tariff, developers rushed to install expensive solar devices, costing consumers trillions of yen that could have been saved by gradual installations made in tandem with cost reductions. In Germany, by contrast, there was a clear incentive for solar developers to reduce cost under its fine-tuned tariff scheme with frequent price adjustments.

Our critics will say that these issues are political, not academic. We feel that this attitude is the source of the problem. Engaged scholarship is a prerequisite for informed policymaking. Scientists and social scientists must do their part.

● **Two Fixes**

Two big changes would go a long way to improving interdisciplinary research in energy and the environment in Japan. Going global is the key, and will pay dividends: Japan would leverage international expertise, and the rest of the world would learn from Japan's experiences.

Globalize the review process. Because of the small number of researchers engaged in interdisciplinary research, the pool of reviewers for academic journals and funding proposals is limited. In policy-relevant interdisciplinary research, particularly in energy and environment, publishers and granting programmes should make parts or all of their review processes international. The connections made could also boost international collaboration.

For strategic research in energy and the environment, funding agencies should require scientists to publish part of their results in international journals even for policy-oriented research, whose target readership obviously prefers Japanese to English. Many papers, although tailored to the policy context of Japan, would appeal to global experts because energy and environmental issues are global. Large-scale programmes of the Science-and-Technology Ministry, and strategic research funds of the Environment Ministry, should take the lead. Because policymakers need deliverables to be communicated in Japanese as well, this will increase the burden on researchers, which should be reflected in their funds.

Globalize research. Strategic, policy-oriented research programmes in Japan should be designed so that they can benefit from international experience and domestic experience can be shared globally. For example, the Collaborative Laboratories for Advanced Decommissioning Science (CLADS), established as a research base for the decommissioning of the Fukushima plant, should be more outward-facing.

Decommissioning involves many disciplines, including nuclear engineering, meteorology and oceanic-risk assessments, ecology and remediation. By soliciting international research proposals, CLADS should involve more researches from elsewhere in Asia, where many countries have nuclear ambitions, including South Korea, India and many southeast Asian countries. Working with overseas scientists, CLADS should publish some outcomes in English.

Another opportunity is Future Earth, a ten-year global sustainability research initiative that puts interdisciplinarity at the forefront alongside stakeholder engagement. For its contribution, Japan should elevate energy research to a key component. Japan has several advanced energy technologies, but to move them into the market at scale requires outside input, particularly when it comes to innovation policy.

- **Better Together**

This year is also the 30th anniversary of the Chernobyl nuclear accident. In Europe, and Germany in particular, that disaster spawned fresh thinking on many fronts. The German book *Risk Society* by Ulrich Beck, published in 1986 soon after the accident, explored how risks from technology and industrialization shape modern society.

The disaster catalysed a transition away from nuclear to renewables, which is now gathering renewed momentum, backed up by interdisciplinary studies on energy transformation. As in Germany in the late 1980s, Japan has seen many fresh attempts to carve out new directions for research, but so far such efforts have been fragmented and scattered, many along disciplinary lines.

Five years on from March 2011, problems abound. Fukushima and the Tohoku areas are yet to recover, and the transition towards renewables has been rocky. Most, if not all, of these issues are fundamentally political and socio-economic. But scientists, social scientists and their funders must engage. Without better connections across disciplines and nations, the science-policy interface cannot improve. The people of Japan deserve better.

(Excerpt from *Nature*, March 3, 2016)

❧ Exercises ☙

▎ Reading Comprehension

Directions: Answer the following questions based on the information from the text.

1. What changes are made in the Japanese energy policy?
2. Why did Japanese scientists lose their credibility and the trust of the public?
3. What is the problem in assessing the safety of nuclear power plants?
4. What does decommissioning mean in the text? What does it involve?
5. How do risks from technology and industrialization shape modern society? Give examples to illustrate it.

Unit 2
Robots and Artificial Intelligence

> **导读**
>
> 本单元主要涉及机器人与人工智能研究。Text A详细阐述了研究人员如何利用信号处理技术、机器学习算法、情感分析工具等来开发情感型机器人。Text B从跨学科的角度介绍了纳米机器人在医疗领域的应用。Text C探讨了人类如何与智能机器人和谐相处的问题，并提出了保障机器人的目标不与人类的目标相冲突的设计原则。

Part I   Intensive Reading

Text A

Robots with Heart

Before we can share our lives with machines, we must teach them to understand and mimic human emotion.

By Pascale Fung

"Sorry, I didn't hear you." This may be the first empathetic utterance by a commercial machine. In the late 1990s the Boston company SpeechWorks International began supplying companies with customer-service software programmed to use this phrase and others. Nearly every call to a customer-service line begins with a conversation with a robot. Hundreds of millions of people carry an intelligent personal assistant around in their pocket.

But machines do not always respond the way we would like them to. Speech-recognition software makes mistakes. Machines often fail to understand intention. They do not get emotion and humor, sarcasm and irony. If in the future we are going to spend more time interacting with machines—whether they are intelligent vacuum cleaners or robotic humanoid nurses—we need them to do more than understand the words we are saying: We need them to get us. We need them, in other words, to "understand" and share human emotion—to possess empathy.

In my laboratory at the Hong Kong University of Science and Technology, we are developing such machines. Empathetic robots can be of great help to society. They will not be mere assistants—they will be companions. They will be friendly and warm, anticipating our physical and emotional needs. They will learn from their interactions with humans. They will make our lives better and our jobs more efficient. They will apologize for their mistakes and ask for permission before proceeding. They will take care of the elderly and teach our children. They might even save your life in critical situations while sacrificing themselves in the process—an act of ultimate empathy.

Some robots that mimic emotion are already on the market—including Pepper, a small humanoid companion built by the French firm Aldebaran Robotics for the Japanese company Softbank Mobile, and Jibo, a six-pound desktop personal-assistant robot designed by a group of engineers that included Roberto Pieraccini, former director of dialog technologies at SpeechWorks. The field of empathetic robotics is still in its steam-engine days, but the tools and algorithms that will dramatically improve these machines are emerging.

● **The Empathy Module**

I became interested in building empathetic robots six years ago, when my research group designed the first Chinese equivalent of Siri. I found it fascinating how naturally users developed emotional reactions to personal-assistant systems—and how frustrated they became when their machines failed to understand what they were trying to communicate. I realized that the key to building machines that could understand human emotion were speech-recognition algorithms like those I had spent my 25-year career developing.

Any intelligent machine is, at its core, a software system consisting of modules, each one a program that performs a single task. An intelligent robot could have one module for processing human speech, one for recognizing objects in images captured by its video camera, and so on. An empathetic robot has a heart, and that heart is a piece of software called the empathy module. An empathy module analyzes facial cues, acoustic markers in speech and the content of speech itself to read human emotion and tell the robot how to respond.

When two people communicate with each other, they automatically use a variety of cues to understand the other person's emotional state—they interpret facial gestures and body language; they perceive changes in tone of voice; they understand the content of speech. Building an empathy module is a matter of identifying those characteristics of human communication that machines can use to recognize emotion and then training algorithms to spot them.

When my research group set out to train machines to detect emotion in speech, we decided to teach machines to recognize fundamental acoustic features of speech in addition to the meaning of the words themselves because this is how humans do it. Our brain detects emotion in a person's voice by paying attention to acoustic cues that signal stress, joy, fear, anger, disgust, and so on. When we are cheerful, we talk faster, and the pitch of our voice rises. When we are stressed, our voices become flat and "dry". Using signal-processing techniques, computers can detect these cues, just as a polygraph picks up blood pressure, pulse and skin conductivity. To detect stress, we used supervised learning to train machine-learning algorithms to recognize sonic cues that correlate with stress.

A brief recording of human speech might contain only a few words, but we can extract vast amounts of signal-processing data from the tone of voice. We started by teaching machines to recognize negative stress (distress) in speech samples from students at my institution. We built the first-ever multilingual corpus of natural stress emotion in English, Mandarin and Cantonese by asking students 12 increasingly stressful questions. By the time we had collected about 10 hours of data, our algorithms could recognize stress accurately 70 percent of the time—remarkably similar to human listeners.

While we were doing this work, another team within my group was training machines to recognize mood in music by analyzing sonic features alone (that is, without paying attention to lyrics). This team started by collecting 5,000 pieces of music from all genres in major European and Asian languages. A few hundred of those pieces had already been classified into 14 mood categories by musicologists.

We electronically extracted some 1,000 fundamental signal attributes from each song—acoustic parameters such as energy, fundamental frequency, harmonics, and so on—and then used the labeled music to train 14 different software "classifiers", each one responsible for determining whether a piece of music belongs to a specific mood. For example, one classifier listens to only happy music, and another listens only for melancholy music. The 14 classifiers work together, building on one another's guesses. If a "happy" classifier mistakenly finds a melancholic song to be happy, then in the next round of relearning, this classifier will be retrained. At each round, the weakest classifier is retrained, and the overall system is boosted. In this manner, the machine listens to many pieces of music and learns which one belongs to which mood. In time, it is able to tell the mood of any piece of music by just listening to the audio, like most of us do.

● Understanding Intent

To understand humor, sarcasm, irony and other high-level communication attributes, a machine will need to do more than recognize emotion from acoustic features. It will also need to understand the underlying meaning of speech and compare the content with the emotion with which it was delivered.

Researchers have been developing advanced speech recognition using data gathered from humans since the 1980s, and today the technology is quite mature. But there is a vast difference between transcribing speech and understanding it.

Think of the chain of cognitive, neurological and muscular events that occurs when one person speaks to another: One person formulates her thoughts, chooses her words and speaks, and then the listener decodes the message. The speech chain between humans and machine goes like this: Speech waves are converted into digital form and then into parameters. Speech-recognition software turns these parameters into words, and a semantic decoder transforms words into meaning.

Unit 2
Robots and Artificial Intelligence

When we began our research on empathetic robots, we realized that algorithms similar to those that extract user sentiment from online comments could help us analyze emotion in speech. These machine-learning algorithms look for telltale cues in the content. Key words such as "sorrow" and "fear" suggest loneliness. Repeated use of telltale colloquial words can reveal that a song is energetic. We also analyze information about the style of speech. Are a person's answers certain and clear or hesitant, peppered with pauses and hedged words? Are the responses elaborate and detailed or short and curt?

In our research on mood recognition in music, we have trained algorithms to mine lyrics for emotional cues. Instead of extracting audio signatures of each piece of music, we pulled strings of words from the song's lyrics and fed them to individual classifiers, each one responsible for determining whether this string of words conveys any of the 14 moods. Such strings of words are called n-grams. In addition to word strings, we also used part-of-speech tags of these words as part of the lyrics "signature" for mood classification. Computers can use n-grams and part-of-speech tags to form statistical approximations of grammatical rules in any language; these rules help programs such as Siri recognize the speech and software such as Google Translate convert text into another language.

Once a machine can understand the content of speech, it can compare that content with the way it is delivered. If a person sighs and says, "I'm so glad I have to work all weekend," an algorithm can detect the mismatch between the emotion cues and the content of the statement and calculate the probability that the speaker is being sarcastic. Similarly, a machine that can understand emotion and speech content can pair that information with other inputs to detect more complex intentions. If someone says, "I'm hungry," a robot can determine the best response based on its location, time of day and the historical preferences of its user, along with other parameters. If the robot and its user are at home and it is almost lunchtime, the robot might know to respond: "Would you like me to make you a sandwich?" If the robot and its user are traveling, the machine might respond: "Would you like for me to look for restaurants?"

• Zara the Supergirl

At the beginning of this year students and postdoctoral researchers in my lab began pulling all our various speech-recognition and emotion-recognition modules together into a prototype empathetic machine we call Zara the Supergirl. It took hundreds of hours of data to train Zara, but today the program runs on a single desktop computer.

When you begin a conversation with Zara, she says, "Please wait while I analyze your face"; Zara's algorithms study images captured by the computer's webcam to determine your gender and ethnicity. She will then guess which language you speak and ask you a few questions in your native tongue. What is your earliest memory? Tell me about your mother. How was your last vacation? Tell me a story with a woman, a dog and a tree. Through this process, based on your facial expressions, the acoustic features of your voice and the content of your responses, Zara will reply in ways that mimic empathy. After five minutes of conversation, Zara will try to guess your personality and ask you about your attitudes toward empathetic machines. This is a way for us to gather feedback from people on their interactions with early empathetic robots.

Zara is a prototype, but because she is based on machine-learning algorithms, she will get "smarter" and more empathetic as she interacts with more people and gathers more data. It would be premature to say that the age of friendly robots has arrived. We are only beginning to develop the most basic tools that emotionally intelligent robots would need. And when Zara's descendants begin arriving on the market, we should not expect them to be perfect. In fact, I have come to believe that focusing on making machines perfectly accurate and efficient misses the point. The important thing is that our machines become more human, even if they are flawed. After all, that is how humans work. If we do this right, empathetic machines will not be the robot overlords that some people fear. They will be our caregivers, our teachers and our friends.

(Excerpt from *Scientific American*, November 2015)

Unit 2
Robots and Artificial Intelligence

Exercises

I. Reading Comprehension

• **Section One**

Directions: Answer the following questions based on the information from the text.

1. What flaws are found in the speech-recognition software?
2. What is the empathy module being developed by the team of Pascale Fung?
3. What is the essential feature of human communication?
4. How did the team make use of music in order to develop empathy module?
5. What did Zara the Supergirl have to do before she could communicate with human beings?

• **Section Two**

Directions: Write an abstract based on the text in no more than 200 words.

Abstract:
Key words:

II. Vocabulary

• **Section One**

Directions: Choose the explanation that is closest in meaning to the underlined part in each sentence.

1. If in the future we are going to spend more time interacting with machines, we need them to do more than understand the words we are saying: We need them to <u>get</u> us.

A. become B. communicate with
   C. understand the point of D. move
2. We need them, in other words, to "understand" and share human emotion—to possess empathy.
   A. sharing sorrow of others
   B. sympathy for another's suffering
   C. giving relief in times of disappointment
   D. ability to share another person's feelings and emotions as if they were your own
3. Some robots that mimic emotion are already on the market.
   A. imitate in order to deter predator or for camouflage
   B. replicate the physiological effects of
   C. pretend to be
   D. imitate in order to entertain or ridicule
4. I became interested in building empathetic robots six years ago, when my research group designed the first Chinese equivalent of Siri.
   A. an exact duplicate
   B. a thing that has the same function in a different place or system
   C. having great or severe effect of something
   D. equal in amount or value
5. We started by teaching machines to recognize negative stress (distress) in speech samples from students at my institution.
   A. extreme anxiety, sorrow or pain
   B. the state of being in extreme danger and needing urgent help
   C. suffering caused by lack of the basic necessities of life
   D. seizure and holding of property as security for payment of a debt
6. Building an empathy module is a matter of identifying those characteristics of human communication that machines can use to recognize emotion and then training algorithms to spot them.
   A. recognize the identity, nationality, etc. of
   B. catch sight of
   C. stain a person's character

Unit 2
Robots and Artificial Intelligence

D. locate an enemy's position from the air

7. Our brain detects emotion in a person's voice by paying attention to acoustic cues that signal stress, joy, fear, anger, disgust, and so on.
   A. reveal the guilt of
   B. discover or perceive the existence or presence of
   C. use an instrument to observe
   D. discover a crime

8. Using signal-processing techniques, computers can detect these cues, just as a polygraph picks up blood pressure, pulse and skin conductivity.
   A. learn or acquire by chance or without effort
   B. become aware of or sensitive to something
   C. detect or receive a signal by means of electronic apparatus
   D. collect something that has been left somewhere

9. It will also need to understand the underlying meaning of speech and compare the content with the emotion with which it was delivered.
   A. bring to a destination
   B. give birth to
   C. utter or recite an opinion, a speech, etc.
   D. hand over

10. These machine-learning algorithms look for telltale cues in the content.
    A. revealing or disclosing unintentionally
    B. revealing information about another's private affairs
    C. mischief-making
    D. story-telling

• Section Two

Directions: There are two or three meanings for each semi-technical term underlined in the following sentences. Choose the correct one according to the context.

1. Any intelligent machine is, at its core, a software system consisting of modules (units or periods of training or education; independent self-contained units of a spacecraft; parts of an electronic system which perform a particular function), each one a program that performs a single task.

2. When we are cheerful, we talk faster, and the pitch (the degree of highness or lowness of

a tone; height, degree, or intensity; the steepness of a slope, especially of a roof) of our voice rises.

3. A brief recording of human speech might contain only a few words, but we can extract (obtain a natural resource from the earth; obtain information from a larger amount or source of information; obtain juice by suction, pressure, distillation, etc.) vast amounts of signal-processing data from the tone of voice.

4. We built the first-ever multilingual corpus (a distinctive structure of some kind in a human or animal body; a large collection of written or spoken texts that is used for language research; a collection of writings, texts, spoken materials, etc.) of natural stress emotion in English, Mandarin and Cantonese by asking students 12 increasingly stressful questions.

5. We electronically extracted some 1,000 fundamental signal attributes from each song—acoustic parameters such as energy (force, vigor, capacity for activity; the capacity of a physical system to do work; means of doing work as provided by the utilization of physical or chemical resources), fundamental frequency, harmonics, and so on.

6. To understand humor, sarcasm, irony and other high-level communication attributes (qualities or features that someone or something has; material objects recognized as appropriate to a person, office, or status), a machine will need to do more than recognize emotion from acoustic features.

7. Think of the chain of cognitive, neurological and muscular events (a thing that happens or takes place, especially one of importance; an item in a sports program; a single occurrence of a process) that occurs when one person speaks to another.

8. One person formulates (prepares according to a formula; expresses a thought, opinion, or idea by using particular words; invents a plan or proposal and thinks about the details carefully) her thoughts, chooses her words and speaks, and then the listener decodes the message.

9. "Sorry, I didn't hear you." This may be the first empathetic utterance (a spoken word, statement, or vocal sound; action of saying or expressing something aloud; an uninterrupted chain of spoken or written language) by a commercial machine.

10. This is a way for us to gather feedback (information about the result of an experiment, etc.; the process in which part of the output of a system is returned to its input in order

to regulate its further output) from people on their interactions with early empathetic robots.

● **Section Three**

Directions: Match the Chinese terms with their English equivalents.

1. 演算法
2. 样机
3. 语气/腔调
4. 信号处理技术
5. 统计近似
6. 情感机器人
7. 声波特征
8. 语义解码器
9. 智能吸尘器
10. 测谎仪
11. 对话框技术
12. 皮肤导电性
13. 声学标记
14. 仿人机器人
15. 语音识别软件

A. speech-recognition software
B. intelligent vacuum cleaner
C. humanoid robot
D. empathetic robot
E. prototype
F. dialog technology
G. algorithm
H. tone of voice
I. signal-processing technique
J. sonic feature
K. acoustic marker
L. semantic decoder
M. statistical approximation
N. skin conductivity
O. polygraph

III. Questions for Discussion

Directions: Work in groups and discuss the following questions.

1. Are you willing to live with humanoid robots? Why or why not?
2. If future robots are more intelligent than human beings in important areas of life, will, human beings become slaves of robots?
3. What special knowledge do you think is needed in the development of empathetic robots?

Part II Extensive Reading

Text B

Launch the Nanobots

Overcoming all the technical challenges may take 20 years or more, but the first steps toward remote-controlled medicine have already been taken.

By Larry Greenemeier

The long-term future envisioned by nanomedicine researchers includes incredibly tiny therapeutic agents that smartly navigate under their own power to a specific target—and only that target—anywhere in the body. On arrival, these self-guided machines may act in any number of ways—from delivering a medicinal payload to providing real-time updates on the status of their disease-fighting progress. Then, having achieved their mission, they will safely biodegrade, leaving little or no trace behind. These so-called nanobots will be made of biocompatible materials, magnetic metals or even filaments of DNA: all materials carefully chosen for their useful properties at the atomic scale, as well as their ability to slip past the body's defenses undisturbed and without triggering any cellular damage.

Although this vision will likely take a decade or two to fulfill, medical researchers have already begun addressing some of the technical problems. One of the biggest challenges is making sure the nano-devices get to their target in the body.

● **Wave Power**

Most drugs on the market today readily float through the body in the bloodstream, either after being injected directly into the blood or, in the case of pills, getting absorbed into the bloodstream from the gastrointestinal tract. But they wind up traveling both to where they are needed and to where they can cause unwanted complications. Sophisticated nanomedicines, in contrast, are being designed to be guided to a tumor or other problem site, where their medicinal payload is released,

Unit 2
Robots and Artificial Intelligence

reducing the chance of side effects.

Magnetic fields and ultrasound waves are the leading candidates for guiding nanomedicines in the near term, says Joseph Wang, chair of nanoengineering and a distinguished professor at the University of California, San Diego. In the magnetic approach, researchers embed nanoparticles of iron oxide or nickel, for example, within a particular medication. They then use an array of permanent magnets positioned outside a mouse or other subject and push or pull the metallic medicine through the body to a selected site by manipulating various magnetic fields. In the ultrasound approach, researchers have directed sound waves at medicine-containing nanobubbles—causing them to burst with enough force that the bubble's cargo can penetrate deep within a targeted tissue or tumor.

Last year medical researchers at Keele University and the University of Nottingham, both in England, added a helpful twist to their magnetic approach in work aimed at healing broken bones. They attached iron oxide nanoparticles to individual stem cells and then injected the preparation into two different experimental environments: fetal chicken femurs and a synthetic bone scaffold made from tissue-engineered collagen hydrogels. Once the stem cells arrived at the break, the researchers used an oscillating external magnetic field to rapidly shift the mechanical stress on the nanoparticles, which in turn transferred the force to the stem cells. This kind of biomechanical stress helped the stem cells to differentiate more effectively into bone. New bone growth occurred in both cases—although overall healing was uneven. Eventually the researchers hope that adding various growth factors to the iron oxide–studded stem cells will make the repair process smoother, says James Henstock, a postdoctoral research associate at Keele's Institute for Science and Technology in Medicine.

- **Autonomous Nanomeds**

The primary drawbacks to the magnetic and acoustic approaches are the need for external guidance—which is cumbersome—and the fact that magnetic fields and ultrasound waves can penetrate only so far into the body. Developing autonomous "micro motors" for the delivery of therapeutic cargo could surmount those problems.

Such micro motors would rely on chemical reactions for propulsion, but toxicity is an issue. For example, oxidizing glucose, a sugar molecule found in the blood, would generate hydrogen peroxide, which could be used as a fuel. But researchers already know that this particular approach would not work in the long run. Hydrogen peroxide corrodes living tissue, and glucose in the body would not produce enough hydrogen peroxide to adequately power micro motors. More promising are efforts to use other naturally occurring substances, such as stomach acid (for applications in the stomach) or water (which is abundant in blood and tissues), as power sources.

Accurate navigation by these self-propelling devices may be an even greater hurdle, however. Just because nanoparticles can move anywhere does not mean that they will necessarily travel exactly where researchers want them to go. Autonomous steering is not yet an option, but a work-around would be to make sure that nanomedicines become active only when they find themselves in the right environment.

To accomplish this trick, researchers have begun creating nanomachines out of synthetic forms of DNA. By ordering the subunits of the molecule so that their electrostatic charges force it to fold in a particular configuration, scientists can engineer the constructs to perform various tasks. For example, some DNA segments may fold themselves into containers that will open and release their contents only when the package comes across a protein important to a disease process or encounters the acidic conditions inside a tumor, says University of Chicago chemistry professor Yamuna Krishnan.

Krishnan and her colleagues envision more advanced, modular entities made of DNA that could be programmed for different tasks, such as imaging or even assembling other nanobots. Yet synthetic DNA is expensive—costing about 100 times more than more traditional materials used to deliver drugs. For now, then, the price discourages drug companies from investing in it as a candidate for treatments, Krishnan says.

Unit 2
Robots and Artificial Intelligence

All of this may be a far cry from building a fleet of smart submarines reminiscent of Proteus in the 1966 film *Fantastic Voyage*. Still, nanobots are finally moving in that direction.

(Excerpt from *Scientific American*, April 2015)

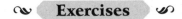 Exercises

I. Translate the following technical terms into English.

1. 纳米医学 _____
2. 生物降解 _____
3. 纳米机器人 _____
4. 生物相容性材料 _____
5. 原子尺度 _____
6. 纳米工程 _____
7. 药用有效载荷 _____
8. 副作用 _____
9. 胃酸 _____
10. 磁场 _____
11. 超声波 _____
12. 干细胞 _____
13. 制剂 _____
14. 生物机械应力 _____
15. 微型马达 _____
16. 静电荷 _____
17. 自主操纵方向 _____
18. 外部引导 _____
19. 化学反应 _____
20. 自动推进式设备 _____

II. Translate the following paragraphs into Chinese.

The primary drawbacks to the magnetic and acoustic approaches are the need for external guidance—which is cumbersome—and the fact that magnetic fields and ultrasound waves can penetrate only so far into the body. Developing autonomous "micro motors" for the delivery of therapeutic cargo could surmount those problems.

Such micro motors would rely on chemical reactions for propulsion, but toxicity is an issue. For example, oxidizing glucose, a sugar molecule found in the blood, would generate hydrogen peroxide, which could be used as a fuel. But researchers already know that this particular approach would not work in the long run. Hydrogen peroxide corrodes living tissue, and glucose in the body would not produce enough hydrogen peroxide to adequately power micro motors. More promising are efforts to use other naturally occurring substances, such as stomach acid (for applications in the stomach) or water (which is abundant in blood and tissues), as power sources.

Accurate navigation by these self-propelling devices may be an even greater hurdle, however. Just because nanoparticles can move anywhere does not mean that they will necessarily travel exactly where researchers want them to go. Autonomous steering is not yet an option, but a work-around would be to make sure that nanomedicines become active only when they find themselves in the right environment.

Text C

Should We Fear Supersmart Robots?

If we're not careful, we could find ourselves at odds with determined, intelligent machines whose objectives conflict with our own.

By Stuart Russell

It is hard to escape the nagging suspicion that creating machines smarter than ourselves might be a problem. After all, if gorillas had accidentally created humans way back when the now endangered primates probably would be wishing they had not done so. But why, specifically, is advanced artificial intelligence a problem?

Hollywood's theory that spontaneously evil machine consciousness will drive armies of killer robots is just silly. The real problem relates to the possibility that AI may become incredibly good at achieving something other than what we really want. In 1960 legendary mathematician Norbert Wiener, who founded the field of cybernetics, put it this way: "If we use, to achieve our purposes, a mechanical agency with whose operation we cannot efficiently interfere..., we had better be quite sure that the purpose put into the machine is the purpose which we really desire."

A machine with a specific purpose has another property, one that we usually associate with living things: a wish to preserve its own existence. For the machine, this trait is not innate, nor is it something introduced by humans; it is a logical consequence of the simple fact that the machine cannot achieve its original purpose if it is dead. So if we send out a robot with the sole directive of fetching coffee, it will have a strong incentive to ensure success by disabling its own off switch or even exterminating anyone who might interfere with its mission. If we are not careful, then, we could face a kind of global chess match against very determined, superintelligent machines whose objectives conflict with our own, with the real world as the chessboard.

The prospect of entering into and losing such a match should concentrate the minds of computer scientists. Some researchers argue that we can seal the machines inside a kind of firewall, using them to answer difficult questions but never allowing them to affect the real world. (Of course, this means giving up on superintelligent robots!) Unfortunately, that plan seems unlikely to work: We have yet to invent a firewall that is secure against ordinary humans, let alone superintelligent machines.

Can we instead tackle Wiener's warning head-on? Can we design AI systems whose goals do not conflict with ours so that we are sure to be happy with the way they behave? This is far from easy, but I believe it is possible if we follow three core principles in designing intelligent systems:

> *The machine's purpose must be to maximize the realization of human values. In particular, the machine has no purpose of its own and no innate desire to protect itself.*
>
> *The machine must be initially uncertain about what those human values are. This turns out to be crucial, and in a way it sidesteps Wiener's problem. The machine may learn more about human values as it goes along, of course, but it may never achieve complete certainty.*
>
> *The machine must be able to learn about human values by observing the choices that we humans make.*

The first two principles may seem counterintuitive, but together they avoid the problem of a robot having a strong incentive to disable its own off switch. The robot is sure it wants to maximize human values, but it also does not know exactly what those are. Now the robot actually benefits from being switched off because it understands that the human will press the off switch to prevent the robot from doing something counter to human values. Thus, the robot has a positive incentive to keep the off switch intact—and this incentive derives directly from its uncertainty about human values.

The third principle borrows from a subdiscipline of AI called inverse reinforcement learning (IRL), which is specifically concerned with learning the values of some entity—whether a human, canine or cockroach—by observing its behavior. By watching a typical human's morning routine, the robot learns about the value of coffee to humans. The field is in its infancy, but already some practical algorithms exist that demonstrate its potential in designing smart machines.

As IRL evolves, it must find ways to cope with the fact that humans are irrational, inconsistent and weak-willed and have limited computational powers, so their actions do not always reflect their values. Also, humans exhibit diverse sets of values, which means that robots must be sensitive to potential conflicts and trade-offs among people. And some humans are just plain evil and should be neither helped nor emulated.

Despite these difficulties, I believe it will be possible for machines to learn enough about human values that they will not pose a threat to our species. Besides directly observing human behavior, machines will be aided by having access to vast amounts of written and filmed information about people doing things. Designing algorithms that can understand this information is much easier than designing superintelligent machines. Also, there are strong economic incentives for robots—and their makers—to understand and acknowledge human values: If one poorly designed domestic robot cooks the cat for dinner, not realizing that its sentimental value outweighs its nutritional value, the domestic robot industry will be out of business.

Solving the safety problem well enough to move forward in AI seems to be feasible but not easy. There are probably decades in which to plan for the arrival of superintelligent machines. But the problem should not be dismissed out of hand, as it has been by some AI researchers. Some argue that humans and machines can coexist as long as they work in teams—yet that is not feasible unless machines share the goals of humans. Others say we can just "switch them off" as if superintelligent machines are too stupid to think of that possibility. Still others think that superintelligent AI will never happen. On September 11, 1933, renowned physicist Ernest Rutherford stated, with utter confidence, "Anyone who expects a source of power in the transformation of these atoms is talking moonshine." On September 12, 1933, physicist Leo Szilard invented the neutron-induced nuclear chain reaction.

(Excerpt from *Scientific American*, June 2016)

Exercises

Reading Comprehension

Directions: Answer the following questions based on the information from the text.

1. What is the real purpose to be put into the machine?
2. What should human beings do if the machine exterminates anyone who might interfere with its mission?
3. How could we ensure that the machines' goals do not conflict with ours so that we are sure to be happy with the way they behave?
4. What is inverse reinforcement learning (IRL)? How could it be achieved?
5. In what ways could machines learn human values?

Unit 3
Education Research

> **导读**
>
> 本单元的文章涉及不同角度的学习科学研究。Text A在认知科学和心理学研究成果的基础上，出人意料地提出，测试一旦做对了，将会是一种非常有效的学习方式。文中有各种实验与分析，启发性很强。在Text B中，研究人员借用医学与经济学领域的研究工具来判断什么是课堂上最好的教学方法，并对一些主流的教育教学理念与方法提出了挑战。Text C详细介绍了慕课的研究方法，对慕课提供的新型数据资源的研究，将极大地促进学习科学的快速发展。

Part I Intensive Reading

Text A

A New Vision for Testing

Too often school assessments heighten anxiety and hinder learning. New research shows how to reverse the trend.

By Annie Murphy Paul

Who was the first American to orbit Earth?
A. Neil Armstrong   B. Yuri Gagarin
C. John Glenn   D. Nikita Khrushchev

In schools across the U.S., multiple-choice questions such as this one provoke anxiety, even dread. Their appearance means it is testing time, and tests are big, important, excruciatingly unpleasant events. But not at Columbia Middle School in Illinois, in the classroom of eighth grade history teacher Patrice Bain. After displaying the question on a smartboard, she pauses as her students enter their responses on numbered devices known as clickers.

"Okay, has everyone put in their answers?" she asks. "Number 19, we're waiting on you!" Hurriedly, 19 punches in a selection, and together Bain and her students look over the class' responses, now displayed at the bottom of the smartboard screen. "Most of you got it—John Glenn—very nice." She chuckles and shakes her head at the answer three of her students have submitted. "Oh, my darlings," says Bain in playful reproach. "Khrushchev was not an astronaut!" Bain moves on to the next question, briskly repeating the process of asking, answering and explaining as she and her students work through the decade of the 1960s.

The banter in Bain's classroom is a world away from the tense standoffs at public schools around the country. Since the enactment of No Child Left Behind in 2002, parents' and teachers' opposition to the law's mandate to test "every child, every year" in grades three through eight has been intensifying. A growing number of parents are withdrawing their children from the annual state tests; the epicenter of the "opt-out" movement may be New York State, where as many as 90 percent of students in some districts reportedly refused to take the year-end examination last spring. Critics of U.S. schools' heavy emphasis on testing charge that the high-stakes assessments inflict anxiety on students and teachers, turning classrooms into test-preparation factories instead of laboratories of genuine, meaningful learning.

In the always polarizing debate over how American students should be educated, testing has become the most controversial issue of all. Yet a crucial piece has been largely missing from the discussion so far. Research in cognitive science and psychology shows that testing, done right, can be an exceptionally effective way to learn. Taking tests, as well as engaging in well-designed activities before and after tests, can produce better recall of facts—and deeper and more complex understanding—than an education without exams. But a testing regime that actively supports learning, in addition to simply assessing, would look very different from the way American schools "do" testing today.

What Bain is doing in her classroom is called retrieval practice. The practice has a well-established base of empirical support in the academic literature, going back almost 100 years—but Bain, unaware of this research, worked out something very similar on her own over the course of a 21-year career in the classroom.

"I've been told I'm a wonderful teacher, which is nice to hear, but at the same time I feel the need to tell people: 'No, it's not me—it's the method,' " says Bain in an interview after her class has ended. "I felt my way into this approach, and I've seen it work such wonders that I want to get up on a mountaintop and shout so everyone can hear me: 'You should be doing this, too!' But it's been hard to persuade other teachers to try it.

Then, eight years ago, she met Mark McDaniel through a mutual acquaintance. McDaniel is a psychology professor at Washington University in St. Louis. He had started to describe to Bain his research on retrieval practice when she broke in with an exclamation. "Patrice said, 'I do that in my classroom! It works!'" McDaniel recalls. He went on to explain to Bain that what he and his colleagues refer to as retrieval practice is, essentially, testing. "We used to call it 'the testing effect' until we got smart and realized that no teacher or parent would want to touch a technique that had the word 'test' in it," McDaniel notes now.

Retrieval practice does not use testing as a tool of assessment. Rather it treats tests as occasions for learning, which makes sense only once we recognize that we have misunderstood the nature of testing. We think of tests as a kind of dipstick that we insert into a student's head, an indicator that tells us how high the level of knowledge has risen in there—when in fact, every time a student calls up knowledge from memory, that memory changes. Its mental representation becomes stronger, more stable and more accessible.

Why would this be? It makes sense considering that we could not possibly remember everything we encounter, says Jeffrey Karpicke, a professor of cognitive psychology at Purdue University. Given that our memory is necessarily selective, the usefulness of a fact or idea makes a sound basis for selection. "Our minds are sensitive to the likelihood that we'll need knowledge at a future time, and if we retrieve a piece of information now, there's a good chance we'll need it again," Karpicke explains.

Studies employing functional magnetic resonance imaging of the brain are beginning to reveal the neural mechanisms behind the testing effect. In the handful of studies that have been conducted so far, scientists have found that calling up information from memory, as compared with simply restudying it, produces higher levels of activity in particular areas of the brain. These brain regions are associated with the so-called consolidation, or stabilization, of memories and with the generation of cues that make memories readily accessible later on.

According to Karpicke, retrieving is the principal way learning happens. "Recalling information we've already stored in memory is a more powerful learning event than storing that information in the first place," he says. "Retrieval is ultimately the process that makes new memories stick." Not only does retrieval practice help students remember the specific information they retrieved, it also improves retention for related information that was not directly tested. Researchers theorize that while sifting through our mind for the particular piece of information we are trying to recollect, we call up associated memories and in so doing strengthen them as well.

Hundreds of studies have demonstrated that retrieval practice is better at improving retention than just about any other method learners could use. To cite one example: In a study published in 2008 by Karpicke and his mentor, Henry Roediger III of Washington University, the authors reported that students who quizzed themselves on vocabulary terms remembered 80 percent of the words later on, whereas students who studied the words by repeatedly reading them over remembered only about a third of the words. Retrieval practice is especially powerful compared with students' most favored study strategies: highlighting and rereading their notes and textbooks, practices that a recent review found to be among the least effective.

And testing does not merely enhance the recall of isolated facts. The process of pulling up information from memory also fosters what researchers call deep learning. Students engaging in deep learning are able to draw inferences from, and make connections among, the facts they know and are able to apply their knowledge in varied contexts (a process learning scientists refer to as transfer). In an article published in 2011 in the journal *Science*, Karpicke and his Purdue colleague Janell Blunt explicitly compared retrieval practice with a study technique known as concept mapping. An activity favored by many teachers as a way to promote deep learning, concept mapping asks students to draw a diagram that depicts the body of knowledge they are learning, with the relations among concepts represented by links among nodes, like roads linking cities on a map.

In their study, Karpicke and Blunt directed groups of undergraduate volunteers—200 in all—to read a passage taken from a science textbook. One group was then asked to create a concept map while referring to the text; another group was asked to recall, from memory, as much information as they could from the text they had just read. On a test given to all the students a week later, the retrieval-practice group was better able to recall the concepts presented in the text than the concept-mapping group. More striking, the former group was also better able to draw inferences and make connections among multiple concepts contained in the text. Overall, Karpicke and Blunt concluded, retrieval practice was about 50 percent more effective at promoting both factual and deep learning.

Transfer—the ability to take knowledge learned in one context and apply it to another—is the ultimate goal of deep learning. In an article published in 2010, psychologist Andrew Butler of University of Texas at Austin demonstrated that retrieval practice promotes transfer better than the conventional approach of studying by rereading. In Butler's experiment, students engaged either in rereading or in retrieval practice after reading a text that pertained to one "knowledge domain"—in this case, bats' use of sound waves to find their way around. A week later the students were asked to transfer what they had learned about bats to a second knowledge domain: the navigational use of sound waves by submarines. Students who had quizzed themselves on the original text about bats were better able to transfer their bat learning to submarines.

Robust though such findings are, they were until recently almost exclusively made in the laboratory, with college students as subjects. McDaniel had long wanted to apply retrieval practice in real-world schools, but gaining access to K–12 classrooms was a challenge. With Bain's help, McDaniel and two of his Washington University colleagues, Roediger and Kathleen McDermott set up a randomized controlled trial at Columbia Middle School that ultimately involved nine teachers and more than 1,400 students. During the course of the experiment, sixth, seventh and eighth graders learned about science and social studies in one of two ways: 1) Material was presented once, then teachers reviewed it with students three times; 2) Material was presented once, and students were quizzed on it three times. When the results of students' regular unit tests were calculated, the difference between the two approaches was clear: Students earned an average grade of C⁺ on material that had been reviewed and A⁻ on material that had been quizzed. On a follow-up test administered eight months later, students still remembered the information they had been quizzed on much better than the information they had reviewed.

"I had always thought of tests as a way to assess—not as a way to learn—so initially I was skeptical," says Andria Matzenbacher, a former teacher at Columbia who now works as an instructional designer. "But I was blown away by the difference retrieval practice made in the students' performance." Bain, for one, was not surprised. "I knew that this method works, but it was good to see it proven scientifically," she says.

Even with the weight of evidence behind them, however, advocates of retrieval practice must still contend with a reflexively negative reaction to testing among many teachers and parents. They also encounter a more thoughtful objection, which goes something like this: If testing is such a great way to learn, why aren't our students doing better?

Marsha Lovett has a ready answer to that question. She admits that American students take a lot of tests. It is what does not happen that causes these tests to fail to function as learning opportunities. Students often receive little information about what they got right and what they got wrong. "That kind of item-by-item feedback is essential to learning, and we're throwing that learning opportunity away," she says. In addition, students are rarely prompted to reflect in a big-picture way on their preparation for, and performance on, the test. "Often students just glance at the grade and then stuff the test away somewhere and never look at it again," Lovett says. "Again, that's a really important learning opportunity that we're letting go to waste."

A few years ago Lovett came up with a way to get students to engage in reflection after a test. She calls it an "exam wrapper". When the instructor hands back a graded test to a student, along with it comes a piece of paper literally wrapped around the test itself. On this paper is a list of questions: a short exercise that students are expected to complete and hand in. The wrapper that Lovett designed for a math exam includes such questions as:

> *Based on the estimates above, what will you do differently in preparing for the next test? For example, will you change your study habits or try to sharpen specific skills? Please be specific. Also, what can we do to help?*

How much time did you spend reviewing with each of the following:
- Reading class notes? _____ minutes
- Reworking old homework problems? _____ minutes
- Working additional problems? _____ minutes
- Reading the book? _____ minutes

Now that you have looked over your exam, estimate the percentage of points you lost due to each of the following:
- _____ % from not understanding a concept
- _____ % from not being careful (i.e., careless mistakes)
- _____ % from not being able to formulate an approach to a problem
- _____ % from other reasons (please specify)

The idea, Lovett says, is to get students thinking about what they did not know or did not understand, why they failed to grasp this information and how they could prepare more effectively in advance of the next test. Lovett has been promoting the use of exam wrappers to the Carnegie Mellon faculty for several years now, and a number of professors, especially in the sciences, have incorporated the technique into their courses.

Does this practice make a difference? In 2013 Lovett published a study of exam wrappers as a chapter in the edited volume *Using Reflection and Metacognition to Improve Student Learning*. It reported that the metacognitive skills of students in classes that used exam wrappers increased more across the semester than those of students in courses that did not employ exam wrappers. In addition, an end-of-semester survey found that among students who were given exam wrappers, more than half cited specific changes they had made in their approach to learning and studying as a result of filling out the wrapper.

Over time, repeated exposure to this testing-feedback loop can motivate students to develop the ability to monitor their own mental processes. Affluent students who receive a topnotch education may acquire this skill as a matter of course, but this capacity is often lacking among low-income students who attend struggling schools—holding out the hopeful possibility that retrieval practice could actually begin to close achievement gaps between the advantaged and the underprivileged. "Repeated testing is a powerful practice that directly enhances learning and thinking skills, and it can be especially helpful to students who start off with a weaker academic background," Gosling says.

However, there lies a dilemma for American public school students, who take an average of 10 standardized tests a year in grades three through eight according to a recent study conducted by the Center for American Progress. Unlike the instructor-written tests given by the teachers and professors profiled here, standardized tests are usually sold to schools by commercial publishing companies. Scores on these tests often arrive weeks or even months after the test is taken. And to maintain the security of test items—and to use the items again on future tests—testing firms do not offer item-by-item feedback, only a rather uninformative numerical score.

There is yet another feature of standardized state tests that prevents them from being used more effectively as occasions for learning. The questions they ask are overwhelmingly of a superficial nature—which leads, almost inevitably, to superficial learning.

If the state tests currently in use in U.S. were themselves assessed on the difficulty and depth of the questions they ask, almost all of them would flunk. That is the conclusion reached by Kun Yuan and Vi-Nhuan Le, both then behavioral scientists at RAND Corporation, a nonprofit think tank. In a report published in 2012 Yuan and Le evaluated the mathematics and English language arts tests offered by 17 states, rating each question on the tests on the cognitive challenge it poses to the test taker. The researchers used a tool called Webb's Depth of Knowledge, which identifies four levels of mental rigor, from DOK1 (simple recall), to DOK2 (application of skills and concepts), through DOK3 (reasoning and inference), and DOK4 (extended planning and investigation).

Most questions on the state tests Yuan and Le examined were at level DOK1 or DOK2. The authors used level DOK4 as their benchmark for questions that measure deeper learning, and by this standard the tests are failing utterly. Only 1 to 6 percent of students were assessed on deeper learning in reading through state tests, Yuan and Le report; 2 to 3 percent were assessed on deeper learning in writing; and 0 percent were assessed on deeper learning in mathematics. "What tests measure matters because what's on the tests tends to drive instruction," observes Linda Darling-Hammond. That is especially true, she notes, when rewards and punishments are attached to the outcomes of the tests, as is the case under the No Child Left Behind law and states' own "accountability" measures.

According to Darling-Hammond, the provisions of No Child Left Behind effectively forced states to employ inexpensive, multiple-choice tests that could be scored by machine—and it is all but impossible, she contends, for such tests to measure deep learning. But other kinds of tests could do so. Darling-Hammond wrote, with her Stanford colleague Frank Adamson, the 2014 book *Beyond the Bubble Test*, which describes a very different vision of assessment: Tests that pose open-ended questions (the answers to which are evaluated by teachers, not machines); that call on students to develop and defend an argument; and that ask test takers to conduct a scientific experiment or construct a research report.

In the 1990s Darling-Hammond points out, some American states had begun to administer such tests; that effort ended with the passage of No Child Left Behind. She acknowledges that the movement toward more sophisticated tests also stalled because of concerns about logistics and cost. Still, assessing students in this way is not a pie-in-the-sky fantasy: Other nations, such as England and Australia, are doing so already. "Their students are performing the work of real scientists and historians, while our students are filling in bubbles," Darling-Hammond says.

She does see some cause for optimism: A new generation of tests are being developed in the U.S. to assess how well students have met the Common Core State

Standards, the set of academic benchmarks in literacy and math that have been adopted by 43 states. Two of these tests—Smarter Balanced and Partnership for Assessment of Readiness for College and Careers (PARCC)—show promise as tests of deep learning, says Darling-Hammond, pointing to a recent evaluation conducted by Joan Herman and Robert Linn. Herman notes that both tests intend to emphasize questions at and above level 2 on Webb's Depth of Knowledge, with at least a third of a student's total possible score coming from questions at DOK3 and DOK4. "PARCC and Smarter Balanced may not go as far as we would have liked," Herman conceded in a blog post last year, but "they are likely to produce a big step forward".

(Excerpt from *Scientific American*, August 2015)

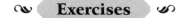

I. Reading Comprehension

● **Section One**

Directions: Locate and write down the findings reported or conclusions drawn by various researchers based on the information from the text.

1. _____

2. _____

3. _____

4. _____

5. _____

6. _____

7. _____

8. _____

● **Section Two**

Directions: Write an abstract based on the text in no more than 200 words.

Abstract:
Key words:

II. Vocabulary

● **Section One**

Directions: Choose the explanation that is closest in meaning to the underlined part in each sentence.

1. Their appearance means it is testing time, and tests are big, important, excruciatingly unpleasant events.

    A. grievously
    B. in a very painful manner
    C. extremely
    D. mentally agonizing

2. The banter in Bain's classroom is a world away from the tense standoffs at public schools around the country.

    A. moving or keeping away
    B. indifference
    C. contests in which the score is tied and the winner is undecided

D. stalemate between two equally matched opponents in a dispute
3. These brain regions are associated with the so-called <u>consolidation</u>, or stabilization, of memories and with the generation of cues that make memories readily accessible later on.

   A. making physically stronger

   B. combining a number of things into a single more effective whole

   C. combining into a single overall account

   D. combining into a solid mass

4. Not only does retrieval practice help students remember the specific information they retrieved, it also improves <u>retention</u> for related information that was not directly tested.

   A. the continued possession, use, or control of something

   B. the fact of keeping something in one's memory

   C. the action of absorbing and continuing to hold a substance

   D. failure to eliminate a substance from the body

5. The questions they ask are overwhelmingly of a superficial nature—which leads, almost inevitably, to <u>superficial</u> learning.

   A. appearing to be real or true only until examined more closely

   B. not thorough, deep, or complete

   C. situated or occurring on the skin

   D. not showing any depth of character

6. To <u>cite</u> one example: In a study published in 2008, the authors reported that students who quizzed themselves on vocabulary terms remembered 80 percent of the words later on, whereas students who studied the words by repeatedly reading them over remembered only about a third of the words.

   A. summon someone to appear in a law court

   B. praise someone for a courageous act in an official dispatch

   C. quote as evidence for or justification of an argument or statement

   D. state it as the official reason for a legal case

7. In a study published in 2008, the authors reported that students who <u>quizzed</u> themselves on vocabulary terms remembered 80 percent of the words later on, whereas students who studied the words by repeatedly reading them over remembered only about a third of the words.

   A. asked someone questions

B. gave a student or class an informal written test

C. made fun of

D. looked curiously at someone through or as if through an eyeglass

8. Lovett has been promoting the use of exam wrappers to the Carnegie Mellon faculty for several years now, and a number of professors, especially in the sciences, have incorporated the technique into their courses.

   A. mixed together different elements

   B. constituted a company or other organization as a legal corporation

   C. combined ingredients into one substance

   D. taken in something as part of a whole

9. Holding out the hopeful possibility that retrieval practice could actually begin to close achievement gaps between the advantaged and the underprivileged.

   A. not enjoying the same standard of living or rights as the majority of people in a society

   B. having special rights, advantages, or immunities

   C. legally protected from being made public

   D. not having special rights, advantages, or immunities

10. A new generation of tests are being developed in the U.S. to assess how well students have met the Common Core State Standards, the set of academic benchmarks in literacy and math that have been adopted by 43 states.

    A. written works with superior or lasting artistic merit

    B. competence or knowledge in a specified area

    C. the ability to read and write

    D. language and literature

• **Section Two**

Directions: There are two or three meanings for each semi-technical term underlined in the following sentences. Choose the correct one according to the context.

1. The epicenter (the point on the earth's surface vertically above the focus of an earthquake; the central point of something, typically a difficult or unpleasant situation) of the "opt-out" movement may be New York State.

2. Critics of U.S. schools' heavy emphasis on testing charge (store electrical energy in a

battery or battery-operated device; formally accuse someone of something or make an assertion that; rush forward in attack) that the high-stakes assessments inflict anxiety on students and teachers.

3. But a testing regime (a system or planned way of doing things, especially one imposed from above; an authoritarian government; a set of rules about food, exercise, or beauty that some people follow in order to stay healthy or attractive) that actively supports learning would look very different from the way American schools "do" testing today.

4. Its mental representation (the action of speaking or acting on behalf of someone; description of someone or something in a particular way; a mental concept regarded as corresponding to a thing perceived) becomes stronger, more stable and more accessible.

5. The authors used level DOK4 as their benchmark (a surveyor's mark cut in a wall and used as a reference point in measuring altitudes; a standard against which things may be compared or assessed; a problem designed to evaluate the performance of a computer system) for questions that measure deeper learning, and by this standard the tests are failing utterly.

6. Transfer (moving from one place to another; application of a skill learned in one situation to a different but similar situation; change to another place, route, or means of transport during a journey) is the ultimate goal of deep learning.

7. Students earned an average grade of C⁺ on material that had been reviewed (read things again and make notes in order to be prepared for the exam; write a critical appraisal of a book, play, film, etc. for publication in a newspaper or magazine; submit a case or sentence for reconsideration by a higher court or authority) and A⁻ on material that had been quizzed.

8. A few years ago Lovett came up with a way to get students to engage in reflection (a thing that is a consequence of or arises from something else; serious thought or consideration about a particular subject) after a test.

9. For example, will you change your study habits or try to sharpen (become better at noticing, thinking, or doing something; raise the pitch of; make one's senses more acute) specific skills?

10. According to Darling-Hammond, the provisions (actions of providing or supplying something for use; conditions or requirements in a legal document; cognitive processes

of thinking about what you will do in the event of something happening) of No Child Left Behind effectively forced states to employ inexpensive, multiple-choice tests that could be scored by machine.

- **Section Three**

Directions: Match the Chinese terms with their English equivalents.

1. 认知心理学　　　　　A. smartboard
2. 深度学习　　　　　　B. clicker
3. 核磁共振成像　　　　C. academic literature
4. 概念构图　　　　　　D. empirical support
5. 实证支持　　　　　　E. retrieval practice
6. 测试—反馈循环　　　F. cognitive psychology
7. 元认知监控　　　　　G. Magnetic Resonance Imaging (MRI)
8. 教学设计师　　　　　H. neural mechanism
9. 表决器/遥控器　　　　I. deep learning
10. 心理过程　　　　　　J. concept mapping
11. 智能白板　　　　　　K. instructional designer
12. 学术文献　　　　　　L. metacognitive monitoring
13. 教育背景　　　　　　M. testing-feedback loop
14. 神经机制　　　　　　N. mental process
15. 回想测试　　　　　　O. academic background

III. Questions for Discussion

Directions: Work in groups and discuss the following questions.

1. What is your opinion about the standardized tests in China?
2. What tests can measure deep learning?
3. How do you usually prepare for a test? Does it work?
4. Does a test allow you to recognize where you are strong and where you need to make improvements?

Unit 3
Education Research

Part II Extensive Reading

Text B

The Science of Learning

Researchers are using tools borrowed from medicine and economics to figure out what works best in the classroom. But the results aren't making it into schools.

By Barbara Kantrowitz

Anna fisher was leading an undergraduate seminar on the subject of attention and distractibility in young children when she noticed that the walls of her classroom were bare. That got her thinking about kindergarten classrooms, which are typically decorated with cheerful posters, multicolored maps, charts and artwork. What effect, she wondered, does all that visual stimulation have on children, who are far more susceptible to distraction than her students at Carnegie Mellon University? Do the decorations affect youngsters' ability to learn?

To find out, Fisher's graduate student Karrie Godwin designed an experiment involving kindergartners at Carnegie Mellon's Children's School. Two groups of 12 kindergartners sat in a room that was alternately decorated with Godwin's purchases or stripped bare and listened to three stories about science in each setting. Researchers videotaped the students and later noted how much each child was paying attention. At the end of the reading, the children were asked questions about what they had heard. Those in the bare classroom were more likely to pay attention and scored higher on comprehension tests.

Hundreds of experiments like Fisher's are part of an effort to bring more rigorous science to U.S. classrooms. The movement started with former president George W. Bush's No Child Left Behind Act and has continued under President Barack Obama. In 2002 the Department of Education established the Institute of Education Sciences (IES) to encourage researchers to pursue what was described as "scientifically valid research", especially randomized controlled trials, which advocates of IES considered the gold standard. The government also created the What Works Clearinghouse to provide a database of results for classroom educators on everything from reviews of particular curricula to evidence-based teaching techniques.

Now researchers are using emerging technology and new methods of data analysis to create experiments that would have been impossible to carry out even 10 years ago. Much of the new research goes beyond the simple metric of standardized tests to study learning in progress. "I am interested in measuring what really matters," said Paulo Blikstein, an assistant professor at the Stanford Graduate School of Education. "We have been developing new technologies and new data-collection methods to capture the process." How well students complete a task is just part of the experiment; researchers also record students' eye gaze, galvanic skin response and exchanges with fellow students, among other things. Blikstein calls this approach "multimodal learning analytics".

The new methodology is already challenging widely held beliefs by finding that teachers cannot be judged solely on the basis of their academic credentials, that classroom size is not always paramount and that students may actually be more engaged if they struggle to complete a classroom assignment. Although these studies have not come up with the "silver bullet" to cure all that ails American schools, the findings are beginning to fill in some blanks in that hugely complex puzzle called education.

- **Looking for Patterns**

Provocative questions are yielding some of the most surprising results. In a series of experiments with middle school and high school students, Blikstein is trying to understand the best ways to teach math and science by going beyond relatively primitive tools like multiple-choice tests to assess students' knowledge. "We bring kids to the lab, and we run studies where we tell them to build some kind of engineering or science project." The researchers put sensors in the lab and sometimes on the kids themselves. Then they collect the data and analyze them to look for patterns. "There are lot of counterintuitive things in how people learn," Blikstein notes. "We like to reveal that an intuition we have is sometimes wrong."

Unit 3
Education Research

"Discovery" learning, in which students discover facts for themselves rather than receiving them directly from an instructor, has been in vogue lately; Blikstein and his colleagues are trying to get at the heart of how much or how little instruction students really need. In one set of studies, they tried to find out whether students learned more about a science topic if they first saw either a lecture or did an exploratory activity. Seeing the lecture first is called "tell and practice", he says. "First you're told, then you practice." Students were divided into two groups: One started with the lecture, and the other started with the exploratory activity. The researchers repeated the experiment in several studies and found fairly consistent results: Students who practiced first performed 25 percent better than students who listened to a lecture first. "The idea here is that if you have a lecture first and you haven't explored the problem by yourself a little bit, you don't even know what questions the lecturing is answering," Blikstein says.

The new tools and methods of data analysis are making education research more efficient and precise. Jordan Matsudaira, a management and policy professor at Cornell University, has helped resurrect an old research tool and has employed it to look at the usefulness of summer school and the effect of funding from Title I, a federal program targeted at schools with a certain percentage of lowincome students. The method, known as regression-discontinuity analysis, compares two groups of students on either side of a particular threshold. For example, in the study on summer school, Matsudaira compared students whose test scores were just above the level that made them eligible for summer school with those who were just below it to see if the extra schooling improved students' test scores. The design is used to mimic randomized controlled trials.

His conclusion: Summer school could be a more cost-effective way of raising test scores than reducing class size.

In the Title I study, Matsudaira compared schools that fell just above the limit required to get the federal funds with those just below it. He found that the money did not make much of a difference in the academic achievement of the students most likely to be affected. But it also illustrated some of the limits of the research design. It is possible that schools with a much higher percentage of poor students might derive a greater benefit from the extra money. It is also possible that schools so close to the threshold would use the money for onetime expenditures rather than longterm investments because they cannot be certain that their population would remain the same and that they would continue to be eligible for the federal aid in the future.

Other researchers are mining data to track the progress of many students over time. Ryan Baker, president of the International Educational Data Mining Society, and his colleagues recently completed a seven-year longitudinal study, funded by the National Science Foundation, looking at log files of how thousands of middle school students used a Web-based math-tutoring program called ASSISTments. The researchers then tracked whether the students went to college and, if they did, how selective the college was and what they majored in to see whether they could make connections between students' use of the software and their later academic achievements.

"Big data allows us to look over long periods, and it allows us to look in very fine detail," Baker says. He and his colleagues were particularly interested in seeing what happened to students who were "gaming" the system—trying to get through a particular set of problems without following all the steps. It turns out that gaming the easier problems was not as harmful as gaming the harder problems. Students who gamed the easier problems could have simply been bored, whereas students who gamed the harder problems might not have understood the material. Baker thinks this kind of information could ultimately help teachers and guidance counselors figure out not only which students are at risk of academic problems but also why they are at risk and what can be done to help them.

- **Building an Evidence Base**

The new studies are helping to build an evidence base that has long been missing in education. Grover Whitehurst, founding director of IES, recalls that when he started in 2002, just after No Child Left Behind took effect, the superintendent of a predominantly minority district asked him to suggest a math curriculum that had been proved effective for his students. "I said, 'There isn't any,'" Whitehurst says. "He couldn't believe that he was being required by law to base everything he did on scientifically based research, and there was none." That superintendent was far from alone, points out Whitehurst. "There was very little research that actually spoke to the needs of policy makers and educators. It was mostly research written by academics and schools of education to be read by academics and schools of education. That was about as far as it went."

Many researchers would disagree with that harsh assessment. Yet the criticism pushed the community to examine and explain its methods and mission. John Easton, current director of IES and a former educational researcher at the University of Chicago, believes the clearinghouse is particularly useful as a way for the government to vet products that school districts might feel pressured to buy. The clearinghouse now houses more than 500 reports that summarize current findings on such topics as math instruction for young children, elementary school writing and helping students with the college application process. It has also reviewed hundreds of thousands of reports to aid in distinguishing the best-quality research from weaker work, including studies on such subjects as the effectiveness of charter schools and merit pay for teachers, which have informed the ongoing debate about these issues.

One of the most important contributions of the government's emphasis on rigorous science, Whitehurst says, has been a dramatic change in the definition of a high-quality teacher. In the past, quality was defined by credentials such as a specific degree or certification. Now, he asserts, "it's about effectiveness in the classroom, measured by observations and measured by the ability of a teacher to increase test scores." Whereas there is still a significant controversy over how to assess an individual teacher's effectiveness, Whitehurst believes that change in approach was driven by the research community, especially economists "who came to this topic because all of sudden there were resources—data resources and research support resources".

Many researchers have complained that the IES's emphasis on randomized controlled trials has disregarded other potentially useful methodologies. Case studies of school districts, for example, could describe learning practices in action the way business schools use case studies of companies. "The current picture is really an ecosystem of methodologies, which makes sense because education is a complex phenomenon if ever there was one—complex in the scientific sense," says Anthony Kelly, a professor of education psychology at George Mason University. Easton says he still believes randomized controlled trials are an important part of that process but not necessarily as "the culminating event". He thinks trials might also be useful early in the process of developing an educational intervention to see whether something is working and worth more investigation.

- **From Lab to Classroom**

Getting this new science into schools remains a challenge. There is also a long-standing barrier between the lab and the classroom. In the past, many researchers felt it was not their job to find real-world applications for their work. And educators for the most part believed that the expertise they gained in the classroom generally trumped anything the researchers could tell them.

The What Works Clearinghouse was supposed to help bridge that gap, but in 2010 the General Accountability Office found that only 42 percent of school districts it surveyed had heard of it. The GAO survey also found that only about 34 percent of districts had accessed the clearinghouse Web site at least once and that even fewer used it frequently. In an updated report in December 2013, the GAO said dissemination remained problematic. The need is more urgent now, with the implementation of the Common Core state standards.

Easton and others have acknowledged the need for a better pipeline to schools. As part of the solution, the clearinghouse has published 18 "practice guides" that lay out what is known about subjects such as teaching students who are learning English or teaching math to young children. Each is compiled by a panel that brings together researchers, teachers and school administrators. The practice guides may also direct future research, says psychology professor Sharon Carver, a member of the early math panel and director of Carnegie Mellon's Children's School. She urges her graduate students to read the guides that relate to their field and look for areas that need more exploration.

Each research question is an attempt to fit in another piece of a very large puzzle. "I don't think you can look at education from the point of view of whether it works or doesn't work, as if it's a lightbulb," says Joseph Merlino, president of the 21st Century Partnership for STEM Education, a nonprofit in suburban Philadelphia. "I don't think human knowledge is like that.... In a mechanical age, we are used to thinking of things mechanically. Does it work? Can you fix it? I don't think you can fix education any more than you can fix your tomato plant. You cultivate it. You nurture it."

Merlino's organization administered a five-year, IES-funded randomized controlled study of the effectiveness of applying four principles of cognitive science to middle school science instruction. A total of 180 schools in Pennsylvania and Arizona were randomly assigned modified or unmodified curricula. One part of the study was based on cognitive science research about how people learn from diagrams. Merlino says the researchers learned that some of the things that graphic artists might put into a diagram to make it jazzy—such as lots of colors—actually distract from learning. The researchers also found that students need instruction in reading diagrams. That is the kind of result that could be integrated into the design of a new textbook. Teachers could also take time to explain the meaning of different symbols in a diagram, such as arrows or cutaways.

Making educators an important part of the research process could also get results into the classroom. Teachers often feel that the expertise they have gained from their experience is ignored and that they instead get a new, supposedly evidence-based curriculum every few years without much explanation of why the new one is so much better than the old. And in the past, researchers have not generally felt that it was their role to explain their work to teachers. That is changing, says Nora Newcombe, a professor of psychology at Temple University and principal investigator of the Spatial Intelligence and Learning Center. "I think people are really waking up to the idea that if you take federal tax dollars, you are supposed to be sharing your knowledge."

The exchange of knowledge can go both ways. In the Pennsylvania and Arizona science curriculum study, teachers were involved in the initial design of the experiments. "They were more like master teachers," Newcombe says. "They taught, and they gave us feedback," she adds. Because the study took place in actual schools rather than a lab, the researchers trained the classroom teachers as the work proceeded.

Other researchers point to the model of Finland, where educational theories, research methodologies and practice are all important parts of teacher education, according to Pasi Sahlberg, who in 2011 wrote *Finnish Lessons*, an account of how the country rebuilt its education system and rose to the top of international math and literacy rankings. In some ways, the comparison to American schools is unfair because Finland is a more homogeneous country. But Newcombe thinks that U.S. teacher training should include the most recent developments in cognitive science. In many teacher education programs, students "are taught a psychology that is not just 10 but more like 40 years out of date", she says. That basic grounding could help teachers assess the importance of new research and find ways to incorporate it into their classrooms. "You can't really write a script for everything that happens in the classroom," Newcombe says. "If you have some principles in your mind for what you do in those on-the-fly moments, you can do a better job."

(Excerpt from *Scientific American*, August 2014)

∽ Exercises ∾

I. Translate the following technical terms into English.

1. 随机对照实验 _____   2. 优质标准 _____
3. 数据分析 _____   4. 衡量标准 _____
5. 标准化考试 _____   6. 眼睛注视 _____
7. 多模态学习分析法 _____   8. 方法论 _____
9. 学术资历 _____   10. 探究式学习 _____
11. 学业成绩 _____   12. 研究设计 _____
13. 纵向研究 _____   14. 大数据 _____
15. 皮肤电流反应 _____   16. 绩效工资 _____
17. 严密科学 _____   18. 个案研究 _____
19. 教育心理学 _____   20. 认知科学 _____

Unit 3
Education Research

■ II. Translate the following paragraph into Chinese.

"Discovery" learning, in which students discover facts for themselves rather than receiving them directly from an instructor, has been in vogue lately; Blikstein and his colleagues at FabLab@School, a network of educational workshops Blikstein created in 2009, are trying to get at the heart of how much or how little instruction students really need. Parents may not like to see their kids frustrated in school, but Blikstein says that "there are levels of frustration and failure that are very productive, are very good ways to learn". In one set of studies, he and his colleagues tried to find out whether students learned more about a science topic if they first saw either a lecture or did an exploratory activity. Seeing the lecture first is called "tell and practice", he says. "First you're told, then you practice." Students were divided into two groups: One started with the lecture, and the other started with the exploratory activity. The researchers repeated the experiment in several studies and found fairly consistent results: Students who practiced first performed 25 percent better than students who listened to a lecture first. "The idea here is that if you have a lecture first and you haven't explored the problem by yourself a little bit, you don't even know what questions the lecturing is answering," Blikstein says.

■ Text C

Rebooting MOOC Research

Improve assessment, data sharing, and experimental design.

By Justin Reich

The chief executive officer of edX, Anant Agarwal, declared that Massive Open Online Courses (MOOCs) should serve as "particle accelerator for learning". MOOCs provide new sources of data and opportunities for large-scale experiments that can advance the science of learning. In the years since MOOCs first attracted widespread attention, new lines of research have begun, but findings from these efforts have had few implications for teaching and learning. Big data sets do not, by virtue of their size, inherently possess answers to interesting questions. For MOOC research to advance the science of learning, researchers, course developers, and other stakeholders must advance the field along three trajectories: from studies of engagement to research about learning, from investigations of individual courses to comparisons across contexts, and from a reliance on post hoc analyses to greater use of multidisciplinary, experimental design.

Clicking or learning? Few MOOC studies make robust claims about student learning, and fewer claim that particular instructional moves caused improved learning. We have terabytes of data about what students clicked and very little understanding of what changed in their heads.

Consider four recent studies conducted on Udacity, Khan Academy, Google Course Builder, and edX. Each study addressed a correlation between measures of student success (such as test scores or course completion) and measures of student activity. All four studies operationalized activity similarly, boiling down the vast data available to a simple, person-level summary variable: number of problems attempted (Udacity), minutes on site (Khan Academy), weekly activity completion (Google), and number of "clicks" per student in the event logs (edX). The complexity of student activity captured by these platforms was lost. Using simple comparisons or regressions, all four concluded there is a positive correlation between student activity and success.

It does not require trillions of event logs to demonstrate that effort is correlated with achievement. As these are observational findings, the causal linkages between doing more and doing better are unclear. Beyond exhorting students to be more active, there are no practical implications for course design. The next generation of MOOC research needs to adopt a wider range of research designs with greater attention to causal factors promoting student learning.

Watching without learning. One reason that early MOOC studies have examined engagement or completion statistics is that most MOOCs do not have assessment structures that support robust inferences about learning. MOOC researchers would, ideally, have assessment data with three characteristics. First, assessments should take place at multiple time points. Pretesting is critical in MOOCs, because heterogeneous registrants include novices and domain experts. Second, assessments should capture multiple dimensions of learning, from procedural to conceptual. Students who earn high grades on quantitative exam questions often show no growth in their conceptual understanding or expert thinking. Finally, courses should include assessments that have been validated by prior research, so comparisons can be made to other settings. With greater attention to assessment in course design, researchers can make stronger claims about what students learn—not just what they do.

Distinguishing between engagement and learning is particularly crucial in voluntary online learning settings, because media that provoke confusion and disequilibrium can be productive for learners. Addressing misconceptions requires addressing the uncomfortable gap between our intuitions and scientific reality. Unfortunately, learners may prefer videos that present material more simply. For instance, students use more positive language to describe instructional videos that present straightforward descriptions of phenomena, even though students learn more from media that directly address misconceptions. Course developers optimizing for engagement statistics can create pleasurable media experiences that keep students watching without necessarily learning.

Rethinking data sharing. Although MOOC researchers have data from thousands of students, few have data from many courses. Student privacy regulations, data protection concerns, and a tendency to hoard data conspire to curtail data sharing. As a result, researchers can examine variation between students but cannot make robust inferences about cross-course differences.

Sharing learner data is no simple matter. Recent efforts to de-identify student data so as to meet privacy requirements demonstrate that the blurring and scrubbing required to protect student anonymity deform data to the point where they are no longer reliable for many forms of scientific inquiry. Enabling a shared science of MOOCs based on open data will require substantial policy changes and new technical innovations in social science data sharing. One policy approach would be to decouple privacy protections from efforts to maintain anonymity, which would allow researchers to share identifiable data in exchange for greater oversight of their data protection regimes. Technical solutions could include regimes based on differential privacy, where institutions would keep student data in a standardized format that allows researchers to query repositories, returning only aggregated results.

Beyond A/B tests. In the absence of shared cross-course data, experimental designs will be central to investigating the efficacy of particular instructional approaches. From the earliest MOOC courses, researchers have implemented "A/B tests" and other experimental designs. These methods are poised to expand as MOOC platforms incorporate authoring tools for randomized assignment of course content.

The most common MOOC experimental interventions have been domain-independent "plug-in" experiments. In one study, students earned virtual "badges" for active participation in a discussion forum. Students randomly received different badge display conditions, some of which caused more forum activity than others. This experiment took place in a Machine Learning class, but it could have been conducted in American Literature or Biology. These domain-independent experiments, often inspired by psychology or behavioral economics, are widely under way in the field. HarvardX, for instance, has recently offered courses with embedded experiments that activate social supports and commitment devices and cause manipulations to increase perceptions of instructor rapport.

The signature advantage of plug-in experiments is that successful interventions to boost motivation, memorization, or other common facets of learning can be adapted to diverse settings. This universality is also a limitation: These studies cannot advance the science of disciplinary learning. They cannot identify how best to address a particular misconception or optimize a specific learning sequence. Boosting motivation in well-designed courses is good, but if a MOOC's overall pedagogical approach is misguided, then plug-in experiments can accelerate participation in ineffective practices. Discipline-based education research to understand domain-specific learning in MOOCs may be prerequisite to effectively leveraging domain-independent research.

There are fewer examples of domain-specific experiments that are "baked-in" to the architecture of MOOCs. Fisher randomly assigned students in his Copyright course to one of two curricula—one based on U.S. case law, the other on global copyright issues—to experimentally assess these approaches. He used final exam scores, student surveys, and teaching assistant feedback to evaluate the curricula and concluded that deep examination of a single copyright regime served students better than a survey of global approaches, providing actionable findings for open online legal education.

Both domain-specific and domain-independent experiments will be important as MOOC research matures, but domain-specific endeavors may require more intentional nurturing. Plug-in experiments fit more easily in the siloed structures of academia, where psychologists and economists can generate interventions to be incorporated in courses developed by others. Domain-specific research requires multidisciplinary teams—content experts, assessment experts, and instructional designers—that are often called for in educational research but remain elusive. More complex MOOC research will require greater institutional support from universities and funding agencies to prosper.

Raising the bar. In a new field, it is appropriate to focus on proof-of-concept demonstrations. For the first MOOC courses, getting basic course materials accessible to millions was an achievement. For the first MOOC researchers, getting data cleaned for any analysis was an achievement. In early efforts, following the path of least resistance to produce results is a wise strategy, but it runs the risk of creating path dependencies.

Using engagement data rather than waiting for learning data, using data from individual courses rather than waiting for shared data, and using simple plug-in experiments versus more complex design research are all sensible design decisions for a young field. Advancing the field, however, will require that researchers tackle obstacles elided by early studies.

These challenges cannot be addressed solely by individual researchers. Improving MOOC research will require collective action from universities, funding agencies, journal editors, conference organizers, and course developers. At many universities that produce MOOCs, there are more faculty eager to teach courses than there are resources to support course production. Universities should prioritize courses that will be designed from the outset to address fundamental questions about teaching and learning in a field. Journal editors and conference organizers should prioritize publication of work conducted jointly across institutions, examining learning outcomes rather than engagement outcomes, and favoring design research and experimental designs over post hoc analyses. Funding agencies should share these priorities, while supporting initiatives—such as new technologies and policies for data sharing—that have potential to transform open science in education and beyond.

(Excerpt from *Science*, January 2, 2015)

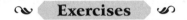

Reading Comprehension

Directions: Answer the following questions based on the information from the text.

1. What should researchers, course developers, and other stakeholders do in order to advance the science of learning?
2. What is the current problem of MOOC studies about student learning?
3. What is the common problem of the four recent studies conducted on Udacity, Khan Academy, Google Course Builder, and edX? What is the possible solution put forward by the author?
4. Why does the author say that sharing learner data is no simple matter? Is there any solution?
5. What should universities, funding agencies, journal editors, conference organizers, and course developers do to meet the challenges MOOC studies face?

Unit 4
Geology and Geoscience

> **导读**
>
> 本单元的文章涉及地质与地球科学研究。Text A 围绕"地球上最古老的岩石"这一科学议题，呈现了两个持不同观点研究团队的争论及其依据。在Text B中，科学家用计算机模型从卫星图像中寻找隐藏的模式，以此生成化石分布图，以帮助科学家缩小搜寻范围。Text C详细讲解了两百年前英国人史密斯如何绘制英国历史上第一张地质图的故事。

Part I　Intensive Reading

Text A

The Oldest Rocks on Earth

One team of scientists thinks ancient rocks discovered in northern Canada give us a window onto the planet's infancy and the birth of life itself. Another team thinks they're not that special.

By Carl Zimmer

Nuvvuagittuq greenstone belt lies in peaceful, roadless isolation along the northeastern edge of Hudson Bay in Canada, more than 20 miles from Inukjuak, the nearest human settlement. From the shoreline, the open ground swells into low hills, some covered by lichens, some scraped bare by Ice Age glaciers. The exposed rocks are beautiful in their stretched and folded complexity. Some are gray and black, shot through with light veins. Others are pinkish, sprinkled with garnets. For most of the year the only visitors here are caribou and mosquitoes.

But this tranquil site is indeed a battleground—a scientific one. For almost a decade rival teams of geologists have traveled to Inukjuak, where they have loaded canoes with camping gear and laboratory equipment and trekked along the coast of the bay to the belt itself. Their goal: to prove just how old the rocks are. One team, headed by University of Colorado geologist Stephen J. Mojzsis, is certain that the age is 3.8 billion years. That is pretty ancient, though not record setting.

Jonathan O'Neil, who leads the competing team at the University of Ottawa, argues that the Nuvvuagittuq rocks formed as long as 4.4 billion years ago. That would make them by far the oldest rocks ever found on Earth. And that is not the least of it. Rocks that old would tell us how the planet's surface formed out of its violent infancy and just how soon after that life emerged—a pivotal chapter in Earth's biography that has so far remained beyond reach.

The first half a billion years of Earth's history—from its formation 4.568 billion years ago to four billion years ago—was a time when water rained down to create the oceans, when the first dry land heaved above the surface of the sea to form continents. It was a time when comets and asteroids❶ crashed into Earth and when a failed planet the size of Mars may have collided with ours, creating the moon from the wreckage. But geologists have very few clues about the timing of these events, such as a few specks of minerals that suggest oceans might have formed before the moon. They find themselves in much the same situation as biographers of ancient Greek philosophers, trying to squeeze as much meaning as they can from scraps of parchment and secondhand stories.

If O'Neil is right and Nuvvuagittuq's rocks are indeed 4.4 billion years old, they will read not like scraps but like entire books. Did plate tectonics start early on, or did Earth mature for hundreds of millions of years before the continents and ocean crust began moving around? What was the chemistry of the youngest oceans and the atmosphere? And how soon did life emerge after Earth formed?

If Mojzsis is right, the earliest chapter in Earth's history will remain shut for now. If O'Neil is right, the rocks of Nuvvuagittuq are among the most precious treasures of geology.

---

❶ Asteroid means a small rocky body orbiting the sun. Large numbers of these, ranging enormously in size, are found between the orbits of Mars and Jupiter.

Unit 4
Geology and Geoscience

• Imprisoned in Stone

Like the rocks that make up much of Earth's crust, the rocks at Nuvvuagittuq generally arose in one of two ways. In some cases, fine particles settled to the bottom of oceans, where they were gradually pressed into layers of sedimentary rock. In other cases, molten magma rose from Earth's mantle, cooling and crystallizing into igneous rock as it ascended.

Only tiny portions of ancient crust in places such as Nuvvuagittuq have remained intact, whereas the rest has vanished. Some rocks were slowly eroded by rain and wind and delivered back to the ocean for new sedimentation. Many others were carried back down under Earth's crust by tectonic plates sinking into the hot mantle, where the rocks melted, their original identity wiped out like an ice cube tossed into a warm pond. Their atoms mixed into the magma and rose again as fresh, young stone.

Rocks on the early Earth were also wiped out by giant asteroids that smashed into the planet and melted large fractions of the crust. About 4.4 billion years ago one collision—called the Giant Impact❶—hurled a huge amount of material into orbit, which became the moon. Given that so much ancient rock was destroyed in one way or another, it is not surprising that samples are rare. That is why the Nuvvuagittuq findings are so prized—and so hotly contested. Just a few other sites around the world have provided samples that are 3.8 billion years old. The oldest is from the tundra of the Northwest Territories❷, dating back 3.92 billion years.

The rarity of early rocks has driven geologists to look for other clues to what the planet was like in its first few hundred million years. Some of those clues have come from tiny crystals called zircons. These rugged, zirconium-based minerals will sometimes form in cooling magma. When the resulting rocks later erode away, some

---

❶ Giant Impact refers to the giant-impact hypothesis, sometimes called Big Splash, which suggests that the Moon formed out of the debris left over from a collision between Earth and an astronomical body the size of Mars, approximately 4.5 billion years ago. Analysis of lunar rocks, published in 2016, suggests that the impact may have been a direct hit, causing a thorough mixing of both parent bodies.

❷ Northwest Territories refers to a large territory in northwestern Canada extending northwards from the 60th parallel and westwards from Hudson Bay to the Rocky Mountains.

of the zircons may remain intact, even as they settle back on the ocean floor and are incorporated into younger sedimentary rocks.

The chemical bonds that make up zircons can trap radioactive atoms such as uranium. The decay of those atoms acts like a clock that geologists can use to measure the age of the zircons. The crystals also trap other chemicals, which can provide a few clues to what Earth was like when they formed. "Zircons are great because they're time capsules❶," Mojzsis says.

In the outback of Australia, geologists have found sedimentary rocks that are sprinkled with immensely old zircons. Some of the zircons (but not the surrounding rock) date back as far as 4.4 billion years, making them the oldest traces of geologic history ever found. Scientists have squeezed remarkable information from these tiny gems since their discovery in 2001. Their structure suggests that the rock in which they originally formed solidified about four miles below the surface. Mojzsis and his colleagues have found chemical fingerprints of water in some of the Australian zircons, too.

The information that scientists can extract from sedimentary zircons is vastly better than nothing, but it is vastly less than they could get from the original rock in which the zircons grew. Rock contains many other minerals, which together can reveal far more about what Earth was like when it formed. "Unless you have the rocks, you don't have the complete story," says Larry Heaman of the University of Alberta, which brings us back to Nuvvuagittuq.

- **Sheer Luck**

In the late 1990s the Quebec government launched a massive geologic expedition to make the first detailed maps of the northern reaches of the province. The region has an onion-like geology, with ancient cores of continental crust surrounded by layers of younger rock. Much of the rock proved to be around 2.8 billion years old. But Pierre Nadeau, then a Ph.D. candidate at Simon Fraser University in British Columbia, brought back a sample that dated back 3.8 billion years. By sheer luck,

---

❶ Time capsule is a container storing a selection of objects chosen as being typical of the present time, buried for discovery in the future.

he had been sent to the Nuvvuagittuq greenstone belt. "Finding these rocks is like having a jewel dropped in my lap," Nadeau's co-worker Ross Stevenson told the BBC in 2002 after they released their results.

Other geologists began to make the long journey to Nuvvuagittuq. Among those pilgrims was O'Neil, who was earning his Ph.D. at McGill University. He was struck by the chemical similarity between the Nuvvuagittuq rocks and 3.8-billion-year-old rocks in Greenland. Perhaps they belonged to the same ancient landmass.

To probe the chemistry, O'Neil teamed up with Carlson from the Carnegie Institution, who is an expert at precision measurements of ancient rocks. The only clear way to determine if a specific stone from Nuvvuagittuq is ancient or not is to date it. To do so, scientists count the levels of radioactive isotopes trapped inside a stone. Radioactive isotopes are variations of atoms that were part of the dusty cloud from which our solar system was born. They became incorporated into solidifying planets and meteorites, and when rocks on Earth crystallized, they became imprisoned inside. As time passed, the isotopes gradually broke down at a regular, clocklike pace. Measuring the levels remaining today reveals a rock's age.

In a Carnegie Institution lab, O'Neil and Carlson tallied up the concentrations of different isotopes. That is when they realized something was very strange about the Nuvvuagittuq samples. Among the isotopes was one known as neodymium 142. It forms from the breakdown of samarium 146. There is no natural samarium 146 left on Earth, because its half-life is short, by some estimates only 68 million years. "It's long gone," Carlson says. "Samarium 146 was present when Earth formed because it was injected by the supernova that started the solar system. But then it decayed away within 500 million years."

Carlson and his colleagues found that different Nuvvuagittuq rocks had different proportions of neodymium 142 and other neodymium isotopes. That variation could have come about only if the rocks formed at a time when there was still samarium 146 on Earth. O'Neil, Carlson and their colleagues compared the proportions to estimate just how long ago the rocks had formed. The number was one none of them had anticipated: 4.28 billion years. To their surprise, they had discovered the oldest rocks on Earth.

"This was totally not what we were expecting to find," O'Neil says. He and his colleagues reported their discovery in 2008. Since then, they have analyzed other samples and now estimate that the Nuvvuagittuq rocks are as old as 4.4 billion years.

- **Old—or Oldest Ever?**

O'Neil and his colleagues first announced their results at a geology conference in Vancouver. Mojzsis can still remember the shock he felt at the news. He had particular reason to be surprised. He was among the few geologists who had traveled to Nuvvuagittuq to follow up on Nadeau's research. Mojzsis and his colleagues had identified a vein of igneous rock that had thrust through the crust after the crust had formed. It turned out to contain zircons. Back home in Colorado, Mojzsis and his colleagues determined that the zircons were 3.75 billion years old—a result that squared nicely with Nadeau's original estimates of 3.8 billion years.

Now O'Neil was standing before Mojzsis and the rest of the scientific community, declaring that the Nuvvuagittuq rocks were half a billion years older. Mojzsis's collaborator Bernard Bourdon of the École Normale Supérieure in Lyon, France, asked for some of O'Neil's samples and tested them again. The measurements of neodymium were correct. Still, "it didn't make sense to me," Mojzsis says.

So, in 2011, Mojzsis and his students returned to Nuvvuagittuq to study the site further. They mapped the terrain and layers of rock around the samples O'Neil had dated. In the rocks that were reportedly 4.4 billion years old, they saw bright green bands of quartzite.

Geologists have seen similar arrangements in much younger formations. They occur when underwater volcanoes spread molten rock across the ocean floor. Sometimes the volcanoes die down, and sediments from the land settle on top of the igneous rocks. Then the volcanoes rev up again, burying the sedimentary rock in a fresh layer of igneous rock. If that was the case in Nuvvuagittuq, then the quartzite had come from sediments from an ancient landmass during one of those volcanic pauses. And if that quartzite had zircons, those zircons would have to be older than the surrounding volcanic rock because they had a much longer history.

"We were on our hands and knees crawling over many outcrops," Mojzsis says.

Unit 4
Geology and Geoscience

After days of hunting, they found two patches of quartzite with zircons—one of which yielded thousands of the tiny minerals. When they brought those zircons back to Colorado, Mojzsis found that they were 3.8 billion years old. Mojzsis's group also approached the question of Nuvvuagittuq's age from other scientific directions. They used another clock to date the rocks, for example, based on the decay of lutetium into hafnium. Once again they came up with 3.8 billion years.

All this evidence has led Mojzsis to a new narrative of Nuvvuagittuq. Around 4.4 billion years ago some molten rock rose toward Earth's surface and turned solid. As it crystallized, it captured some short-lived radioactive samarium 146 that still existed in the early Earth. But then the ancient crust was pulled back down into the mantle. The material heated up to the point where it was no longer rock, but all of it did not get mixed into the surrounding mantle. A bit of it remained a distinct blob with its own peculiar levels of neodymium. Finally, 600 million years later, volcanic activity pushed the material back to the surface, creating rock that incorporated some of the ancient blob, along with the blob's 4.4-billion-year-old signature. "That melt itself can have a memory of a previous existence," Mojzsis says. As a result, a rock that is only 3.8 billion years ago can appear to be 4.4 billion years old.

- **Zircon Mystery Explained**

Mojzsis and his colleagues have been presenting these results at geologic conferences, sometimes in the same sessions where O'Neil is presenting the opposing view—that the rocks formed 4.4 billion years ago and simply have remained there in Earth's crust ever since. O'Neil's team has returned to Nuvvuagittuq, building up its collection of ancient rocks from 10 to about 50. None of the new data have clashed with the original estimation for the age of the site. O'Neil also rejects the evidence that Mojzsis and his colleagues have used to argue that Nuvvuagittuq is only 3.8 billion years old. "We have strong disagreement about the geology of the region," O'Neil says.

Take the quartzite layer where Mojzsis found his zircons. In formations as old as those of Nuvvuagittuq, it is not simple to identify what kind of rock makes up a formation, because it has been deformed so much over billions of years. O'Neil does not think the quartzite band is quartzite at all. Instead, he argues, it is a vein of magma that pushed itself into the ancient rock 3.8 billion years ago. The age of its zircons thus has no bearing on the age of the surrounding rock. "There's nothing bizarre or unusual" about his own rocks, O'Neil says. "They're just really old."

Heaman, himself an expert on old rocks, thinks that O'Neil and his colleagues have made a good case. "I think their evidence is compelling," he says. "They've done their due diligence." But Heaman also thinks that some uncertainty will endure until scientists can find another way to date the rocks. It is possible that a few minerals are lurking in the contested Nuvvuagittuq rocks that contain uranium and lead. That combination is the most reliable way to tell ancient time because scientists have vast experience with it.

- **When Life Formed**

If the nuvvuagittuq rocks are indeed 4.4 billion years old, O'Neil believes they have the potential to open a wide window on the early Earth because they would have formed shortly after the Giant Impact. The Australian zircons were also forming at that time, several miles down into the mantle. But O'Neil argues that the Nuvvuagittuq rocks formed on the surface. "The geochemistry of these rocks really looks like an ocean floor," he says.

If that is true, it confirms that Earth acquired an ocean not long after the Giant Impact. O'Neil also finds that the chemistry of the rocks is remarkably similar to seafloor rocks that formed much more recently. That would suggest that when the world's oceans first arose, they were not drastically different than they are today. O'Neil even believes that the rocks show signs of plate tectonics, suggesting that this process started very early in the planet's history.

There is an even more exciting prospect if the Nuvvuagittuq rocks formed on the ocean floor 4.4 billion years ago: They could shed light on the origin of life. Right now the fossil trail runs cold at 3.5 billion years ago. In rocks younger than that, scientists find preserved bacteria. In rocks older than that, they have found none.

But fossils are not the only traces that life can leave behind. As bacteria feed on carbon, they can alter the balance of carbon isotopes in their environment, and that imbalance can be preserved in rocks that form at the time. Some researchers have claimed that the 3.8-billion-year-old rocks from Greenland carry that imbalance, a signature of life.

That still leaves no evidence of life for the first 700 million years of Earth's existence. Scientists thus cannot say whether life got a quick start on Earth shortly after the planet formed or if it was delayed for hundreds of millions of years. They also have yet to figure out where life began on the planet. Some researchers have suggested that biological molecules emerged in deserts or tidal pools. Others have contended that deep-sea hydrothermal vents were the original nurseries.

If the Nuvvuagittuq rocks formed on the ocean floor 4.4 billion years ago, they are the perfect material to study to tackle these big questions. O'Neil hopes to collaborate with researchers to see whether the rocks could have formed at hydrothermal vents. Finding the earliest traces of life is an obsession of Mojzsis, too, but he will not be looking in Nuvvuagittuq for them. "I'll spend the rest of my career pursuing that jabberwocky❶," he says.

Mojzsis does see an important benefit arising from his disagreement with O'Neil and others, however. As they spar, they are developing better methods for dating old rocks in general. Future generations of geologists who venture into the remote corners of the world and bring back enigmatic samples will be able to use those methods to finally lift the curtain back on the early Earth. And on that point, at least, O'Neil and Mojzsis agree. "All these small enclaves❷ of old rocks are probably all over the place," O'Neil says. "They're just really easy to miss."

(Excerpt from *Scientific American*, March 2014)

---

❶ Jabberwocky means invented or meaningless language; nonsense. It comes from the title of a nonsense poem in Lewis Carroll's *Through the Looking Glass* (1871).
❷ Enclave refers to a portion of territory within or surrounded by a larger territory whose inhabitants are culturally or ethnically distinct; (Figurative) a place or group that is different in character from those surrounding it.

I. Reading Comprehension

● **Section One**

Directions: Answer the following questions based on the information from the text.

1. How did geologists date the rocks?
2. How many "clocks" did geologists use to reveal the age of rocks?
3. What is the disagreement between the team headed by Mojzsis' and O'Neil's team?
4. What is Heaman's opinion about O'Neil's finding?
5. How signifcant would it be if the Nuvvuagittuq rocks are indeed 4.4 billion years old?

● **Section Two**

Directions: Write an abstract based on the text in no more than 200 words.

Abstract:

Key words:

II. Vocabulary

● **Section One**

Directions: Choose the explanation that is closest in meaning to the underlined part in each sentence.

1. Rocks that old would tell us how the planet's surface formed out of its violent infancy and just how soon after that life emerged—a pivotal <u>chapter</u> in Earth's biography that has so far remained beyond reach.

A. a main division of a book B. a section of a treaty
C. a distinctive period in history D. a series or sequence

2. Rocks that old would tell us how the planet's surface formed out of its violent infancy and just how soon after that life emerged—a pivotal chapter in Earth's <u>biography</u> that has so far remained beyond reach.

    A. an account of someone's life written by someone else
    B. the course of a person's life
    C. the writing as a branch of literature
    D. an account in biographical form of an organization, society, animal, etc.

3. They find themselves in much the same situation as biographers of ancient Greek philosophers, trying to <u>squeeze</u> as much meaning as they can from scraps of parchment and secondhand stories.

    A. firmly press with one's fingers
    B. obtain something with difficulty
    C. have a damage effect on something
    D. extract liquid from something by compressing it firmly

4. The chemical <u>bonds</u> that make up zircons can trap radioactive atoms such as uranium.

    A. ropes or chains used to hold someone prisoner
    B. agreements or promises with legal force
    C. strong forces of attraction holding atoms together in a molecule or crystal
    D. things used to tie something or to link things together

5. The chemical bonds that make up zircons can <u>trap</u> radioactive atoms such as uranium.

    A. hold fast and prevent from moving
    B. catch from a trap
    C. catch a criminal
    D. trick someone so that they do or say something that they do not want to

6. The age of its zircons thus has no <u>bearing</u> on the age of the surrounding rock.

    A. relation or relevance
    B. a person's way of standing or moving
    C. the ability to tolerate something bad
    D. the direction or position of something relative to a fixed point

7. Heaman, himself an expert on old rocks, thinks that O'Neil and his colleagues have made a good case. "I think their evidence is <u>compelling</u>," he says.

   A. evoking interest, attention, or admiration in a powerfully irresistible way

   B. inspiring conviction, not able to be refuted

   C. overwhelming, not able to be resisted

   D. driving or forcing

8. There is an even more exciting prospect if the Nuvvuagittuq rocks formed on the ocean floor 4.4 billion years ago: They could <u>shed light on</u> the origin of life.

   A. understand something after prolonged thought or doubt

   B. take into consideration

   C. make widely known or evident

   D. help to explain by providing further information about it

9. If the Nuvvuagittuq rocks formed on the ocean floor 4.4 billion years ago, they are the perfect material to study to <u>tackle</u> these big questions.

   A. initiate discussion with someone about a disputed or sensitive issue

   B. make determined efforts to deal with a problem or difficult task

   C. try to take the ball from an opponent by intercepting them

   D. attack someone and fight them

10. As they <u>spar</u>, they are developing better methods for dating old rocks in general.

    A. engage in argument, typically of a kind that is prolonged or repeated but not violent

    B. fight with the feet or spurs in a game-cock

    C. make the motions of boxing without landing heavy blows as a form of training

    D. shut close

● **Section Two**

Directions: There are two or three meanings for each semi-technical term underlined in the following sentences. Choose the correct one according to the context.

1. The exposed rocks are beautiful in their stretched and folded complexity. Some are gray and black, shot through with light <u>veins</u> (blood vessels; streaks or stripes of a different color in wood, marble, etc.; distinctive qualities, styles, or tendencies).

Unit 4
Geology and Geoscience

2. In other cases, molten magma rose from Earth's mantle, cooling and <u>crystallizing</u> (forming or causing to form crystals; making definite and clear) into igneous rock as it ascended.
3. Some rocks were slowly eroded by rain and wind and delivered back to the ocean for new <u>sedimentation</u> (matter that settles to the bottom of a liquid; particulate matter that is carried by water or wind and deposited on the surface of the seabed and may in time become consolidated into rock).
4. The <u>decay</u> (organic matter rots or decomposes through the action of bacteria and fungi; radioactive substance or particle undergoes change to a different form by emitting radiation; decline in quality, power, or vigor) of those atoms acts like a clock that geologists can use to measure the age of the zircons.
5. Their structure suggests that the rock in which they originally formed <u>solidified</u> (changed from a liquid or fluid into a solid and became hard; made stronger, reinforced) about four miles below the surface.
6. In the late 1990s the Quebec government launched a massive geologic expedition to make the first detailed maps of the northern <u>reaches</u> (the distance to which someone can stretch out their hand; the extent or range of application, effect, or influence; a continuous extent of land or water, especially a stretch of river between two bends, or the part of a canal between locks) of the province.
7. The only clear way to determine if a specific stone from Nuvvuagittuq is ancient or not is to <u>date</u> (go out with someone in whom one is romantically interested; establish or ascertain the date of an object or event; originate at a particular time or have existed since) it.
8. Radioactive isotopes are <u>variations</u> (something a little different from others of the same type; an instance of change; the angle between magnetic north and true north) of atoms that were part of the dusty cloud from which our solar system was born.
9. In a Carnegie Institution lab, O'Neil and Carlson tallied up the <u>concentrations</u> (close gatherings of people or things; the relative amounts of a particular substance contained within a solution or mixture or in a particular volume of space; the actions of strengthening solution by the removal of water or other diluting agent) of different isotopes.

10. Among the isotopes was one known as neodymium 142. It forms from the breakdown (the chemical or physical decomposition of something; a failure of a relationship; an explanatory analysis, especially of statistics) of samarium 146.

● **Section Three**

Directions: Match the Chinese terms with their English equivalents.

1. 海洋地壳   A. Ice Age
2. 沉积岩   B. camping gear
3. 熔融岩浆   C. plate tectonics
4. 地幔   D. ocean crust
5. 冰河时代   E. sedimentary rock
6. 野营装备   F. molten magma
7. 板块构造论   G. mantle
8. 化学指纹图谱   H. igneous rock
9. 放射性同位素   I. chemical fingerprint
10. 地球化学   J. radioactive isotope
11. 大陆地壳   K. half-life
12. 生物分子   L. continental crust
13. 水下热液喷口   M. geochemistry
14. 火成岩   N. biological molecule
15. 半衰期   O. hydrothermal vent

III. Questions for Discussion

Directions: Work in groups and discuss the following questions.

1. What do you think we can learn from geoscience?
2. What is the ultimate goal of studying Earth?
3. As a major in some universities, geoscience is not a choice for many students. What is the possible reason behind it?

Unit 4
Geology and Geoscience

Part II Extensive Reading

Text B

Fossil GPS

By Robert L. Anemone and Charles W. Emerson

On a broiling day in July 2009, a caravan of four-wheel-drive vehicles traveled a faint, two-track dirt road in southwestern Wyoming's Great Divide Basin. The expedition was headed for an area known as Salt Sage Draw in search of buried treasure: fossils dating to between 55 million and 50 million years ago, at the start of the Eocene epoch❶, when the ancestors of many modern orders of mammals were beginning to replace the more archaic mammals that had existed during the earlier Paleocene epoch❷. One of us (Anemone) had been leading field crews of anthropologists, paleontologists and geologists to the basin since 1994, and Salt Sage Draw had proved a fruitful hunting ground over the years, yielding fossils at several localities. Yet this time I was having trouble finding the site. It dawned on me that the road we were on was not the one we had used in previous years. My error would turn out to be very fortunate indeed.

As the tracks began to disappear in the sagebrush and tall grass, I stopped the caravan and walked a way to see if I could spot the road ahead. Rounding a small hill, I noticed an extensive bed of sandstone in the near distance and the elusive road right alongside it. Because sandstone in the Great Divide Basin and many other sedimentary basins in the American West often harbors fossils, I decided to spend some time searching these deposits before we resumed our trip to Salt Sage Draw.

---

❶ Eocene epoch: of, relating to, or denoting the second epoch of the Tertiary period, between the Palaeocene and Oligocene epochs. The Eocene epoch lasted from 56.5 to 35.4 million years ago. It was a time of rising temperature, and there was an abundance of mammals, including the first horses, bats, and whales.

❷ Paleocene epoch: of, relating to, or denoting the earliest epoch of the Tertiary period, between the Cretaceous (白垩纪) period and the Eocene epoch. The Paleocene epoch lasted from 65.0 to 56.5 million years ago. It was a time of sudden diversification among the mammals, probably as a result of the mass extinctions (notably of the dinosaurs) which occurred at the end of the Cretaceous period.

After about an hour of systematically scanning the rock on hands and knees, my then graduate students Tim Held and Justin Gish shouted that they had found a couple of nice mammal jaws. I eagerly joined them. Fossil jaws with teeth are prized because they contain enough information to identify the kind of animal they came from, even in the absence of other parts of the skeleton, and because they reveal what the animal ate.

What came next can only be described as every paleontologist's dream. My students had located a fossil "hotspot". But this was no ordinary hotspot with a handful of jaws or a few dozen teeth and bones eroding out of the sandstone. Rather they had found an extraordinary trove from which we have now collected nearly 500 well-preserved jaws and several thousand teeth and bones from more than 20 different fossil mammal species that lived here approximately 50 million years ago. We call the spot "Tim's Confession", and today it remains not only our best site in the Great Divide Basin but also one of the richest caches of early Eocene mammals in the entire American West.

Mine is hardly the first team to make a major fossil discovery more or less by accident. The history of paleontology is littered with such tales of serendipity. In fact, the ways that vertebrate paleontologists attempt to locate productive fossil sites have not changed much since the earliest days of our science. Like the 19th-century pioneers of our field, we use geologic and topographic evidence to determine where we might have the best chance of finding fossils eroding out of ancient sediments. But beyond that, whether we hit pay dirt❶ is still largely a matter of luck, and more often than not the hard work of looking for fossils goes unrewarded.

Our experience at Tim's Confession got me thinking about whether there might be a better way to determine where my field crew should spend its efforts searching for new fossil sites. We knew that the fossils we were interested in occur in sandstone dating to between 55 million and 50 million years ago, and we knew where in the basin some of these sedimentary layers were exposed and thus suitable

---

❶ When miners would dig for gold, they would say that they had hit pay dirt when they found dirt that contained gold. Hit pay dirt means that you found something that turned out to be profitable or beneficial.

for exploration. But although that information helped to narrow our search somewhat, it still left thousands of square kilometers of ground to cover and plenty of opportunities to come up empty-handed.

Then one night in camp, an idea began to germinate. Out in the field, kilometers away from the nearest source of light pollution, we often noticed satellites passing overhead. I wondered whether we could somehow combine our expert knowledge of the local geology, topography and paleontology of the Great Divide Basin with a satellite's view of the entire 10,000-square-kilometer area to, in essence, map its probable fossil hotspots. Perhaps satellites could "see" features of the land invisible to the naked eye that could help us find more sandstone outcrops and distinguish those that contain accessible fossils from those that do not.

- **Eyes in the Sky**

Other paleontologists, of course, have speculated about whether satellite imagery might improve our ability to find fossils in the field. As a specialist in the fossil record of primate and human evolution, I knew that in the 1990s, Berhane Asfaw of the Rift Valley Research Service and his colleagues had used such images to identify rock exposures in Ethiopia that might yield fossils of human ancestors. At around the same time, Richard Stucky of the Denver Museum of Nature & Science demonstrated that different rock units in the fossil-rich Wind River Basin in central Wyoming could be distinguished and mapped based on analysis of satellite imagery of the region. Both these projects involved collaborations between paleontologists and remote-sensing specialists from NASA and proved the value of such cross-disciplinary efforts. But I wondered if there was a way to tease more information out of the satellite images and thus better focus our search.

I turned to a geographer, the other author of this article (Emerson), and the two of us soon sketched out a plan. We would obtain freely available images of the basin from the Landsat 7 satellite and its so-called Enhanced Thematic Mapper Plus sensor, which detects radiation reflected or emitted from the earth's surface in wavelengths spanning the electromagnetic spectrum—from the blue to the infrared—and represents it in eight discrete spectral bands. The bands can be used to distinguish soil from vegetation, for example, or to map mineral deposits. Then we would develop a method that would allow us to characterize the radiation profiles of known productive fossil localities in the Great Divide Basin based on satellite imagery and see if they shared a telltale spectral signature. If so, we could search the entire Great Divide Basin from our computers to locate new sites that share this spectral signature and thus have a high probability of bearing fossils. We could then visit those places in person and exhaustively search them for fossils to test the model.

Determining whether our known fossil sites shared a distinctive spectral signature was no small task, because for each site we had to assess the combination of values in six bands of the electromagnetic spectrum provided by the Landsat data. Our problem was essentially one of pattern recognition in multiple dimensions, something that humans do not do particularly well but that computers excel at. So we enlisted a so-called artificial neural network—a computational model capable of learning complex patterns.

Our artificial neural network revealed that the basin's known fossil sites do indeed share a spectral signature, and it was able to easily tell these sandstone localities apart from other types of ground cover, such as wetlands and sand dunes. But the model had its limitations. Neural networks are analytical "black boxes", meaning they can distinguish patterns, but they do not reveal the actual factors that allow different patterns to be distinguished. So whereas our neural network could easily and accurately distinguish fossil localities from wetlands or sand dunes, it could not tell us how the spectral signatures of different land covers actually differed in the six bands of the Landsat data—information that could conceivably help us conduct a more targeted search. Another limitation of the neural network approach is that it is based entirely on the analysis of individual pixels. The problem is that the area of an individual Landsat pixel, which measures 225 square meters, does not necessarily correspond to the size of a fossil locality: Some localities are larger than an individual pixel; some are smaller. Thus, the neural network's predictions about the location and extent of potential fossil sites do not always match up with reality.

To overcome these constraints, we needed to be able to analyze multiple adjacent

Unit 4
Geology and Geoscience

and spectrally similar pixels and to statistically describe the distinctive spectral signature of the entire area, whether it was a fossil site or a forest. We turned to a technique known as geographical object-based image analysis and to commercially available, high-resolution satellite imagery in which individual pixels were less than one meter in diameter. Unlike an artificial neural network, this approach allows satellite images to be segmented into image objects—that is, groups of spectrally homogeneous pixels—that can then be characterized by statistical parameters such as mean or median brightness or texture. These image objects more closely match points of interest on the ground, such as fossil sites or stands of forest. Using this image-analysis technique, we were able to develop an independent set of predictions about where to find fossils.

• Moment of Truth

Both our predictive models yielded maps of the Great Divide Basin that pinpointed unexplored areas whose spectral signatures most closely resembled those of the known localities. Although the models exhibited a good degree of overlap in their predictions, they also diverged in some cases. We chose to focus on those places that both models identified as high-priority potential sites. Maps in hand, we headed out to Wyoming during the summers of 2012 and 2013 to see if our models would lead us to new fossil caches in the Great Divide Basin. Gratifyingly, they did exactly that.

The artificial neural network model turns out to be extremely efficient at identifying sandstone deposits, which are almost always worth exploring because so many of the ones in this basin contain fossil vertebrates. One of the first sandstones it led us to in July 2012 yielded a dozen fossils of characteristic Eocene mammals, including the five-toed horse *Hyracotherium*, the early primate *Cantius* and several other creatures belonging to an extinct group of hoofed mammals known as the *Condylarthra*. The neural network also guided us to several spots that yielded aquatic fossil vertebrates, including fish, crocodiles and turtles.

Our geographical object-based image analysis model took us to new sites, too. After a slow start in which the first three or four places the model pointed us to give up no fossils, we moved to the northern part of the Great Divide Basin, near a place called Freighter Gap, for a week of intensive "ground truthing" of our new technique. Graduate student Bryan Bommersbach, who a week before had led us on a long hike to a place that was entirely barren of fossils (we dubbed it "Bryan's Folly"), took the lead in choosing which areas to survey based on the model's predictions. Almost immediately, we began to find bones at many of these locations. We searched for remains at 31 separate places on the landscape that our model indicated were spectrally similar to known localities and found vertebrate fossils at 25 of these places, which is a much higher success rate than is typical when surveying without the help of a predictive map. Mammal fossils emerged from 10 of these localities, one of which dates to the latest part of the Paleocene—an extremely rare find.

We have every reason to believe that predictive models akin to the ones we developed will work in regions other than the Great Divide Basin. In fact, they should work virtually anywhere in the world. In a conservative test of this approach, we used the neural network we developed for the Great Divide Basin to predict the locations of fossil-bearing sedimentary deposits in the nearby Bison Basin, which is known to harbor Paleocene mammal fossils. Encouragingly, our neural network predicted the three most productive fossil localities known in the Bison Basin. Thus, a field crew exploring this vast area for the first time using our predictive model would have had a far better chance of discovering these sites than a crew using traditional survey methods.

Our trial runs in 2012 and 2013 in Wyoming showed that the use of satellite imagery in combination with geospatial predictive models greatly increased the effectiveness of our fieldwork, helping us to find more fossils in less time. But we still have more to do. We are now focused on refining our models to better characterize and differentiate the spectral signature of productive localities. We are convinced that with these tools we can put the future of paleontological exploration on a more secure and scientific footing and reduce the role of serendipity in finding important fossils. Achieving that goal will be well worth the effort required.

(Excerpt from *Scientific American*, May 2014)

Unit 4
Geology and Geoscience

∞ Exercises ∞

I. Translate the following technical terms into English.

1. 古生物学家 _____
2. 地质学家 _____
3. 沉积盆地 _____
4. 野外调查队 _____
5. 地形学 _____
6. 沉积层 _____
7. 遥感专家 _____
8. 资源探测卫星7号 _____
9. 电磁频谱 _____
10. 频谱带 _____
11. 矿床 _____
12. 辐射轮廓 _____
13. 光谱特征 _____
14. 模式识别 _____
15. 人工神经网络 _____
16. 计算模型 _____
17. 高分辨率卫星图像 _____
18. 化石沉积层 _____
19. 图像分析技术 _____
20. 地理空间预测模型 _____

II. Translate the following paragraphs into Chinese.

Determining whether our known fossil sites shared a distinctive spectral signature was no small task, because for each site we had to assess the combination of values in six bands of the electromagnetic spectrum provided by the Landsat data. Our problem was essentially one of pattern recognition in multiple dimensions, something that humans do not do particularly well but that computers excel at. So we enlisted a so-called artificial neural network—a computational model capable of learning complex patterns.

Our artificial neural network revealed that the basin's known fossil sites do indeed share a spectral signature, and it was able to easily tell these sandstone localities apart from other types of ground cover, such as wetlands and sand dunes. But the model had its limitations. Neural networks, by their very nature, are analytical "black boxes", meaning they can distinguish patterns, but they do not reveal the actual factors that allow different patterns to be distinguished. So whereas our neural network could easily and accurately distinguish fossil localities from wetlands or sand dunes, it could not tell us how the spectral signatures of different land covers actually differed in the six bands of the Landsat data—information that could conceivably help us conduct a more targeted search. Another limitation of the neural network approach is that it is based entirely on the analysis of individual pixels. The problem is that the area of an individual Landsat pixel, which measures 225 square meters, does not necessarily correspond to the size of a fossil locality: Some localities are larger than an individual pixel; some are smaller. Thus, the neural network's predictions about the location and extent of potential fossil sites (or a certain type of ground cover, for that matter) do not always match up with reality.

Text C

The Birth of the Geological Map

Two hundred years ago, William Smith published the first geological map of a country.

By Tom Sharpe

In 1815, William Smith, an English canal surveyor and land drainer, provided the young science of geology with the first true geological map of an entire country. Two hundred years on, Smith's map has become an icon of Earth science, and the basic principles he developed and applied are still used in interpreting rock sequences and making geological maps.

Smith's map is remarkable for many reasons. It was ambitious in its scale and scope, covering the whole of England, Wales, and southern Scotland, an area of more than 175,000 km^2. It was constructed by the application of Smith's own discovery that the strata of southern Britain are arranged in a regular sequence and that each rock layer contains distinctive and diagnostic fossils. It was big, measuring about 2.6 m by 1.8 m. It was colorful, with each color carefully chosen for its similarity to that of the rocks it represented. And it was the work of a single individual with a limited rural education, the son of a village blacksmith, working independently outside of the developing structure of the recently formed Geological Society of London.

Smith's was not the first map to show the distribution of different rock types; such "mineralogical" maps had been around for some time. But it was the first to show the rocks of a whole country in a way that indicated the sequence of the strata and with a key in an informative, stratigraphic order, from oldest rocks to youngest. The real innovation, however, was the clever use of a darker watercolor tone to indicate the base of each stratum, fading out toward the top of the bed to give an impression of three dimensions. At a glance, not only is the sequence of tilted beds, layered one on top of the other, readily apparent, but the viewer can immediately envisage how the beds continue underground as they dip below overlying strata.

Smith's great map, *A Delineation of the Strata of England and Wales with Part of Scotland*, was first exhibited in London in early 1815. Its subtitle—*Exhibiting the Collieries and Mines, the Marshes and Fen Lands Originally Overflowed by the Sea, and the Varieties of Soil According to the Variations in the Substrata*—reflected where Smith saw the value of his map: It was intended to be a practical tool for mineral exploration, land drainage, and agriculture.

Smith had begun his career as a trainee surveyor at a time of land enclosure to create large private estates, before being sent to survey coal mines in Somerset and working as surveyor for the route of the proposed Somerset Coal Canal near Bath in the west of England. During the course of this work, Smith noticed that the two branches of the canal in adjacent valleys passed through the same sequence of strata and that the different rocks were always in the same order and arrangement, tilted gently toward the southeast. In addition, he recognized that each bed of rock contained distinctive fossils. This discovery of the value of fossils in correlating strata was a fundamental breakthrough for the new science of geology. In continental Europe, unknown to Smith, others had been developing the same idea, mainly in small areas, but it was Smith who recognized that the principle of diagnostic fossils could be applied over a wide area.

Smith's discovery had immediate practical application. These were the early days of the Industrial Revolution in Britain, when coal was in huge demand. Land owners could make a fortune from coal on their properties, but many searching for it were being misled: Dark gray mudstones similar to those associated with coal rocks also occur far above and below the coal beds in much older and much younger strata. Fossils allowed Smith to distinguish between these different mudstone sequences and to say with confidence whether coal was likely to be found.

By 1799, Smith was in a position to describe a sequence of 23 strata around Bath and to draw up a rudimentary geological map of the area. Working as an itinerant surveyor and drainage engineer, Smith was able to validate his ideas on the regularity of strata and the use of fossils as he traveled the length and breadth of England and Wales. Within 2 years, he was in a position to draft outline geological maps, tracing some of the strata across England to the east coast. Much of the geology of southern and eastern England is a southeasterly tilted sequence of fossil-bearing sedimentary rocks, lacking the complications of igneous intrusion, severe folding, or metamorphism found in much of western and northern Britain. Nonetheless, the preparation of his great map was to take another 14 years.

Smith's map and his ideas on strata and the use of fossils did not receive universal acclaim from his contemporaries. George Bellas Greenough and the gentlemen of the Geological Society were unconvinced by Smith's approach and, by the time Smith's map was published, they had begun work on their own map of the geology of England and Wales. This collaboration by the Society's members around the country soon ran into difficulties, which were resolved—on the recommendation of John Farey, one of Smith's pupils and friends—by an application of Smithian principles. Smith's ideas worked. Smith's map evolved during the years of its production and was modified as new information became available to him. As a result, at least five different versions are recognized today.

Making a geological map remains a first step to understanding regional geology and in the search for raw materials and hydrocarbons. The location and development of resources such as rare earth minerals in places like Greenland requires a geological map at the outset. In hydrocarbon exploration today, seismic surveys extend the mapping into the subsurface in three or even four dimensions. Fossils are as essential in correlating rocks as they were in Smith's time and are important in the identification of strata in the oil industry today.

Unit 4
Geology and Geoscience

Geological maps have, since the time of NASA's Apollo program, extended beyond the terrestrial, as exemplified by a new geological map of Mars. The paper map is now being replaced by the digital, and the same technology allows us to view Smith's map differently by rendering it in three dimensions. In his own lifetime, Smith was hailed the "Father of English Geology", but he can equally be regarded as the Father of Stratigraphy and of the modern geological map.

(Excerpt from *Science*, January 16, 2015)

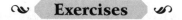

Reading Comprehension

Directions: Answer the following questions based on the information from the text.

1. For what reason is Smith's geological map remarkable?
2. What is the innovation in and the value of Smith's geological map?
3. What is considered to be a fundamental breakthrough for the new science of geology? Explain it in detail.
4. What was the immediate practical application of Smith's discovery (of diagnostic fossils)?
5. What is the significance of a geological map?

Unit 5
GM Technology

> **导读**
>
> 本单元的文章均涉及生物（转基因）技术的应用和争议。Text A详细讲解了最新的基因编辑工具在农产品中的应用，并以解决美国的蘑菇褐变问题为例，说明了基因编辑技术在农业领域应用的优势，部分涉及CRISPR。Text B提供了另一个转基因技术应用的具体案例。在这个案例中，研究人员试图使用转基因技术拯救一度辉煌的美国板栗树。Text C详细阐述了研究人员如何改造水稻基因，使之适应盐渍地，产出正常的大米，并可做成香甜可口的米饭。同样的技术亦可用于改造其他农作物，使之适应盐渍地。

Part I Intensive Reading

Text A

Editing the Mushroom

A powerful new gene-editing tool is sweeping agriculture. It could transform the debate over genetic modification.

By Stephen S. Hall

The hundred or so farmers crowding the ballroom of the Mendenhall Inn in Chester County, Pennsylvania, might not have had a background in gene editing, but they knew mushrooms. These local growers produce a staggering 1.1 million pounds of mushrooms on average every day, which is one reason Pennsylvania dominates the annual $1.2-billion U.S. market. Some of the mushrooms they produce, however, turn brown and decay on store shelves. Mushrooms are so sensitive to physical insult that even careful "one-touch" picking and packing can activate an enzyme that hastens their decay.

At a continuing education seminar on mushrooms last fall, a biologist named Yinong Yang took the podium to deliver news of a possible solution for the browning problem. Yang, a cheerfully polite professor of plant pathology at Pennsylvania State University, is not an expert in the field. But he edited the genome of *Agaricus bisporus*, the most popular dinner-table mushroom in the Western world, using a new tool called CRISPR.

The mushroom farmers in the audience had probably never heard of CRISPR, but they understood it was a big deal. And they understood the enormous commercial implications when Yang showed them photographs comparing brown, decayed mushrooms with pristine white CRISPR-engineered *A. bisporus*, the all-purpose strain that annually accounts for more than 900 million pounds of white button, cremini and portobello mushrooms.

In its brief three years as a science story, CRISPR has already generated more fascinating subplots than a Dickens novel. It is a revolutionary research tool with dramatic medical implications, thorny bioethical conundrums, an awkward patent spat and billion-dollar commercial implications for medicine and agriculture. Academic laboratories and biotech companies are chasing novel treatments for diseases such as sickle-cell anemia and beta-thalassemia. The prospect of using CRISPR to repair embryos or permanently edit our DNA (a process known as human germ-line modification) has sparked fevered talk of "improving" the human species and calls for international moratoriums.

The CRISPR revolution may be having its most profound—and least publicized—effect in agriculture. By the fall of 2015 about 50 scientific papers had been published reporting uses of CRISPR in gene-edited plants, and there are preliminary signs that the U.S. Department of Agriculture does not think all gene-edited crops require the same regulatory attention as "traditional" genetically modified organisms, or GMOs. With that regulatory door even slightly ajar, companies are racing to get gene-edited crops into the fields and, ultimately, into the food supply.

The transformative aspect of CRISPR lies in its unprecedented precision. CRISPR allows you to knock out any gene or, with a little more effort, to add a desirable trait by inserting a gene in a specific place in a genome. It also enables scientists to sidestep, in many cases, the controversial techniques of inserting DNA from other species into plants; these "transgenic" crops, such as the Monsanto-made corn and soybeans that are resistant to the herbicide Roundup, have aroused particular ire in GMO critics and led to public distrust of the technology. Yet some scientists are optimistic that CRISPR crops are so fundamentally different that they will change the tenor of the debate over GMO foods.

Will consumers agree? Or will they see CRISPR crops as the latest iteration of Frankenfood❶—a genetic distortion of nature in which foreign (and agribusiness-friendly) DNA is muscled into a species, with unpredictable health or environmental consequences? Because CRISPR is only now being applied to food crops, the question has not yet surfaced for the public, but it will soon.

● **"Wow, That's the One!"**

The telltale sign of any transformational technology is how quickly researchers apply it to their own scientific problems. By that standard, CRISPR ranks among the most powerful additions to biology's tool kit in the past half a century. The gene-edited mushroom is a case in point.

Yinong Yang dabbled with some primitive gene-editing enzymes in the mid-1990s as a graduate student at the University of Florida and later at the University of Arkansas. He vividly remembers opening the August 17, 2012, issue of *Science*, which contained a paper describing CRISPR's gene-editing potential. "Wow," he thought. "That's the one!" Within days he was hatching plans to improve traits in rice and potato plants through gene editing. His lab published its first CRISPR paper

---

❶ Frankenfood, derogatory genetically modified food. Frankenfood, like Frankenstein, is fiction. Frankenstein is a character in the novel *Frankenstein*, or the *Modern Prometheus* (1818) by Mary Shelley. Baron Frankenstein is a scientist who creates and brings to life a manlike monster which eventually turns on him and destroys him; Frankenstein is not the name of the monster itself, as is often assumed. Now "Frankenstein" is often used to refer to a thing that becomes terrifying or destructive to its maker.

Unit 5
GM Technology

in the summer of 2013.

He was not alone. Plant scientists jumped on CRISPR as soon as the technique was published. Chinese scientists, who quickly embraced the technology, shocked the agricultural community in 2014 when they showed how CRISPR could be used to make bread wheat resistant to a long-standing scourge, powdery mildew.

The gene-editing revolution had begun before the arrival of CRISPR, however. Without the hoopla of CRISPR, agricultural scientists have used TALENs to produce gene-edited plants that have already been grown in fields in North and South America. Calyxt, for example, has created two strains of soybean modified to produce a healthier oil, with levels of monosaturated fats comparable to olive and canola oils. And the company has gene-edited a potato strain to prevent the accumulation of certain sugars during cold storage, reducing the bitter taste associated with storage, as well as the amount of acrylamide, a suspected carcinogen, produced when potatoes are fried.

Because these genetic modifications did not involve the introduction of any foreign genes, the USDA's Animal and Plant Health Inspection Service (APHIS) decided last year that the crops do not need to be regulated as GMOs. "The USDA has given regulatory clearance to plant a potato variety and two soybean varieties, so the potato and one of the soybean varieties are in the field this year," Voytas told me last October. "They basically considered these as just standard plants, as if they were generated by chemical mutagens or gamma rays or some nonregulated technology. The fact that we got regulatory clearance and can go almost immediately from the greenhouse to the field is a big plus. It allows us to really accelerate product development."

Animal scientists have also jumped on the gene-editing bandwagon. Researchers at the small Minnesota-based biotech firm Recombinetics have genetically blocked the biological signal that governs the growth of horns in Holstein cows, the workhorse of the dairy industry. They accomplished this by using gene editing to replicate a mutation that naturally occurs in Angus beef cattle, which do not grow horns. Ag scientists tout this application of gene editing as a more humane form of farming because it spares male Holstein cows from a gruesome procedure during which dairy farmers physically gouge out and then cauterize developing horns (the procedure is done to protect both dairy cattle and dairy farmers from injury).

The mushroom story took a decisive turn in October 2013, when a Penn State alum named David Carroll, president of Giorgi Mushroom, popped into Yang's lab, wondering if new gene-editing techniques could be used to improve mushrooms. Emboldened by the power of CRISPR to create highly precise mutations, Yang replied, "What kind of trait do you want?" Carroll suggested anti-browning, and Yang immediately agreed to try it. Yang knew exactly which gene he wanted to target. Biologists had previously identified a family of six genes, each of which encodes an enzyme that causes browning. Four of the so-called browning genes churn out that enzyme in the fruiting body of mushrooms, and Yang thought that if he could shut down one of them through a gene-editing mutation, he might slow the rate of browning.

The brilliant ease of CRISPR derives from the fact that it is straightforward for biologists to customize a molecular tool—a "construct"—that creates such mutations. Like a utility knife that combines a compass, scissors and vise, these tools excel at two tasks: homing in on a very specific stretch of DNA and then cutting it. The homing is accomplished by a small piece of nucleic acid called the guide RNA, which is designed to mirror the DNA sequence in the target area and attached to it using the unique and specific attraction of DNA base pairs made famous by James Watson and Francis Crick (where As grab onto Ts and Cs grab onto Gs). If you make a piece of guide RNA that is 20 letters long, it will find its mirror sequence of DNA—with GPS-like precision—amid the string of 30 million letters that spell out the Agaricus mushroom genome. The cutting is then accomplished by the Cas9 enzyme, originally isolated from bacterial cultures in yogurt, which rides in on the back of the guide RNA.

Once gene editors cut DNA at the desired spot, they let nature perform the dirty work of mutation. Any time the double helix of DNA is cut, the cell notices the wound and sets out to repair the break. These repairs are not perfect, however, which

Unit 5
GM Technology

is exactly what makes CRISPR so powerful at creating mutations. During the repair process, a few letters of DNA usually get deleted; because a cell's protein-making machinery reads DNA in three-letter "words", deleting a couple of letters subverts the entire text and essentially inactivates the gene by creating what is known as a reading frame shift. That is precisely what happened with the gene-edited mushroom. In Yang's work, a tiny deletion of DNA inactivated one of the enzymes that promote browning—a mutation that Yang and his colleagues confirmed with DNA analysis.

It is fast, cheap and easy. It took about two months of lab work to create the anti-browning mushroom; Yang's demeanor suggested that the work was routine, if not ridiculously easy. And it was remarkably inexpensive. The trickiest step, making the guide RNA and its scaffolding, cost a couple of hundred dollars. The biggest cost is manpower: Xiangling Shen, a postdoctoral fellow in Yang's lab, worked on the project part-time. "If you don't consider manpower, it probably cost less than $10,000," Yang says. In the world of agricultural biotech, that is chump change❶.

And that doesn't begin to hint at the potentially game-changing thrift of CRISPR in the regulatory arena. Last October, Yang gave an informal presentation of the mushroom work to federal regulators at the USDA's APHIS, which decides if genetically modified food crops fall under government regulatory control (in short, whether they are considered a GMO); he came away from the meeting convinced that USDA regulators did not believe the CRISPR mushroom would require special or extended regulatory review. If true, that may be the most important way CRISPR is cheaper: Voytas has estimated that the regulatory review process can cost up to $35 million and take up to five and a half years.

Another advantage of the mushroom as a proof of principle for CRISPR in agriculture is the speed at which fungi grow: From spawn to maturity, mushrooms take about five weeks, and they can be grown year-round in windowless, climate-controlled facilities known as mushroom houses. The gene-edited soybeans and potatoes created by Calyxt, in contrast, take months to field-test.

One of the long-standing fears about genetic modification is the specter

---

❶ Chump change is informal, which means a small or insignificant amount of money and was originally Black English in the 1960s.

of unintended consequences. In the world of biotech foods, this usually means unexpected toxins or allergens making modified foods unhealthy or a genetically modified crop running amok and devastating the local ecology. CRISPR is even making people such as John Pecchia think about unintended economic consequences. One of two mushroom professors at Penn State, Pecchia took some of Yang's starter culture and grew up the first batch of gene-edited mushrooms in the spring of 2015. Standing outside a room where a steamy, fetid mix of mushroom compost was brewing at 80 degrees Celsius, he notes that a mushroom with a longer shelf life might result in smaller demand from stores and also enable unexpected competition. "You could open up the borders to foreign mushroom imports," he adds, "so it's a double-edged sword."

In the tortuous path of genetically modified foods to market, here is one more paradox to chew on. No one knows what the gene-edited mushroom tastes like. They've been steamed and boiled, but not for eating purposes. Every mushroom created so far has been destroyed after Yang conducted browning tests. Once proof of principle has been established, Pecchia says, "We just steam them away."

● Transgene-Free Modification

Will the public steam, sauté or otherwise welcome gene-edited food into their kitchens and onto their plates? That may be the central question in the most intriguing chapter in the CRISPR food story, which coincides with a crucial juncture in the tumultuous, 30-year debate over genetically modified crops.

When Yang described his mushroom project to the Pennsylvania farmers—and to officials at the USDA last October—he used a telltale phrase to describe his procedure: "transgene-free genetic modification". The phrase is a carefully crafted attempt to distinguish the new, high-precision gene-editing techniques like CRISPR from earlier agricultural biotech, where foreign DNA (transgenes) were added to a plant species. Indeed, the acronym "GEO" (for gene-edited organism) has begun to crop up as an alternative to "GMO" or "GM".

The reframing is as much philosophical as semantic, and it is unfolding as the

Obama administration is overhauling the system by which the government reviews genetically modified crops and foods. Known as the Coordinated Framework for Regulation of Biotechnology, this regulatory process, which has not been updated since 1992, defines roles for the USDA, the Food and Drug Administration, and the Environmental Protection Agency. The power of CRISPR has added urgency to the regulatory rethink, and scientists are using the opportunity to revisit a very old question: What exactly does "genetically modified" mean? Voytas, whose track record of publications and patents in gene-edited food crops makes him a sort of editor in chief of small agricultural biotechs in the U.S., answered with a grim little laugh when asked that question: "The GM term is a tricky one."

What's so tricky about it? Most critics of biotech food argue that any form of genetic modification is just that, genetic modification, bringing with it the possibility of unintended mutations or alterations that could pose risks to human health or the environment. Scientists such as Voytas and Yang reply that all forms of plant breeding, dating all the way back to the creation of bread wheat by Neolithic farmers 3,000 years ago, involve genetic modification and that the use of traditional breeding techniques is not a biologically benign process.

Before the era of recombinant DNA in the 1970s, which allowed first-generation agricultural biotech, plant breeders typically resorted to brute-force methods (X-rays, gamma rays or powerful chemicals) to alter the DNA of plants. Despite this blunderbuss approach, some of these random, man-made mutations modified genes in a way that produced desirable agricultural traits: higher yields, or more shapely fruit, or an ability to grow in adverse conditions such as drought. These beneficial mutations could then be combined with beneficial traits in other strains but only by crossing—or mating—the plants. That type of crossbreeding takes a lot of time (often five to 10 years), but at least it is "natural".

But it is also very disruptive. Any time DNA from two different individuals comes together during reproduction, whether in humans or plants, the DNA gets scrambled in a process known as chromosomal reassortment. Spontaneous mutations can occur in each generation, and millions of base pairs of DNA can be transferred when breeders select for a desired trait. Moreover, the desirable trait often drags along with it an undesirable trait on the same piece of DNA during the process of breeding. On the basis of several recent findings on the genetics of rice plants, some biologists hypothesize that domestication has inadvertently introduced "silent" detrimental mutations as well as obvious beneficial traits.

Although CRISPR is more precise than traditional breeding, the technique is not infallible. The precision cutting tool sometimes cuts an unintended region, and the frequency of these "off-target" cuts has raised safety concerns. Feng Zhang of the Broad Institute has published several refinements in the CRISPR system that improve specificity and reduce off-target hits.

The ease and relative thrift of CRISPR have also allowed academic labs and small biotechs back into a game that has historically been monopolized by big agribusinesses. Only deep-pocketed companies could afford to run the costly regulatory gauntlet in the beginning, and to date, almost every crop modification created by genetic engineering was done to enhance the economics of food production for farmers or companies, and they were not very food-centric, but now, the multinational corporations that have dominated the field for the past decade and a half do not have a glowing record in terms of innovation beyond traits for pesticide and herbicide resistance.

The new players have brought a different kind of innovation to agriculture. Voytas, for example, argues that the precision of gene editing is allowing biotech scientists to target consumers by creating healthier, safer foods. Voytas and his colleague Caixia Gao of the Chinese Academy of Sciences have pointed out that plants have many "antinutritionals": noxious self-defense substances or outright toxins that could be gene-edited away to improve nutritional and taste traits. Calyxt's gene-edited potato, for example, reduces a bitter taste trait associated with cold storage of the tubers.

Like any powerful new technology, CRISPR has inspired some agricultural dreamers to envision almost science-fiction scenarios for the future of farming. Michael Palmgren, a plant biologist at the University of Copenhagen, has proposed that scientists can use the new gene-editing techniques to "rewild" food plants, that is, to resurrect traits that have been lost during generations of agricultural breeding. Attempts at rewilding are already under way but with a twist. Rather than restoring lost wild traits to domestic breeds, Voytas says his University of Minnesota lab is attempting what he calls "molecular domestication": transferring agriculturally desirable genes from existing hybrids back into wild species that are hardier and more adaptable, such as the ancestral form of corn, and potatoes. "It's usually only a handful of critical changes that occurred—five, six or seven genes—that allowed a weedy species to become desirable, such as changes in fruit size or corn ear number, those sorts of things," Voytas says. Rather than crossing the wild varieties with the domesticated strains, he says, "Maybe we can just go in and treat those genes and domesticate the wild variety."

There are early signs that gene editing, including CRISPR, may also enjoy a speedier regulatory path. So far U.S. regulators appear to view at least some gene-edited crops as different from transgenic GMO crops. To companies, this suggests that U.S. authorities view the new techniques as fundamentally distinct from transgenic methods; to critics, it suggests a regulatory loophole that companies are exploiting. Yang's mushrooms may be the first CRISPR food considered by the USDA.

The day of food-market reckoning for gene-edited crops may not be too far off. How will the public respond? Kuzma predicts that people who have historically opposed genetic modification will not be drinking CRISPR Kool-Aid anytime soon. She is more concerned about the need to revamp the overall regulatory structure and bring more voices into the review process, at an "inflection point[1]" at which more and more gene-edited foods are wending their way to the marketplace.

(Excerpt from *Scientific American*, March 2016)

---

[1] Inflection point, or point of inflection, which refers in mathematics to a point of a curve where a change in the direction of curvature occurs, is also often used in business meaning a time of significant change in a situation, a turning point.

EST Reading

Exercises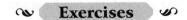

I. Reading Comprehension

● **Section One**

Directions: Answer the following questions based on the information from the text.

1. What did Yinong Yang do to slow the rate of browning with gene-editing tools?
2. What is the point of biotech food critics and that of the proponents?
3. What is the difference between genetic modification and transgene-free modification?
4. What is the difference between traditional GM corporations and small bio-techs (new players) in this field?
5. What is the attitude of the U.S. government regulators toward gene-edited crops?

● **Section Two**

Directions: Write an abstract based on the text in no more than 200 words.

Abstract:
Key words:

II. Vocabulary

● **Section One**

Directions: Choose the explanation that is closest in meaning to the underlined part in each sentence.

1. Mushrooms are so sensitive to physical <u>insult</u> that even careful "one-touch" picking and packing can activate an enzyme that hastens their decay.

Unit 5
GM Technology

A. treating with disrespect or scornful abuse

B. disrespectful abusive remark or action

C. an event which causes damage to a tissue or organ

D. a thing so worthless or contemptible as to be offensive

2. It is a revolutionary research tool with dramatic medical implications, <u>thorny</u> bioethical conundrums, an awkward patent spat and floating over it all, billion-dollar commercial implications for medicine and agriculture.

   A. having many thorns or thorn bushes

   B. causing distress, difficulty, or trouble

   C. continuously uneasy, especially in fear of being detected

   D. a source of continual annoyance

3. Academic laboratories and biotech companies are chasing <u>novel</u> treatments for diseases such as sickle-cell anemia and beta-thalassemia.

   A. interestingly new or unusual

   B. a long written story about imaginary people and events

   C. literary genre represented by novel works

   D. an amendment to an existing statute

4. Ag scientists tout this application of gene editing as a more humane form of farming because it <u>spares</u> male Holstein cows from a gruesome procedure during which dairy farmers physically gouge out and then cauterize developing horns.

   A. gives something of which one has enough to someone

   B. tries to satisfy one's own comfort or needs

   C. is unwanted or not needed and therefore available for use

   D. refrains from killing, injuring, or distressing

5. Yet some scientists are optimistic that CRISPR crops are so fundamentally different that they will change the <u>tenor</u> of the debate over GMO foods.

   A. the general meaning, sense, or content of something

   B. the highest singing voice of the ordinary adult male range

   C. a settled or prevailing course of a person's life or habits

   D. the actual wording of a document

6. The fact that we got regulatory clearance and can go almost immediately from the greenhouse to the field is a big plus.
   A. mathematic operation of addition
   B. advantage
   C. together with
   D. above zero
7. In Yang's work, a tiny deletion of DNA inactivated one of the enzymes that promote browning—a mutation that Yang and his colleagues confirmed with DNA analysis.
   A. had no chemical or biological effect on something
   B. did not engage in any physical activity
   C. did not work or made inoperative
   D. did not exhibit symptoms
8. Like a utility knife that combines a compass, scissors and vise, these tools excel at two tasks: homing in on a very specific stretch of DNA and then cutting it.
   A. becoming fully realized by someone
   B. returning by instinct to its territory after leaving it
   C. focusing attention on
   D. being aimed at a target with great accuracy
9. Only deep-pocketed companies could afford to run the costly regulatory gauntlet in the beginning.
   A. having a source of substantial wealth
   B. dishonestly receiving money for oneself
   C. a narrow sack used as a measure for trading agricultural produce
   D. spending money like water
10. It's usually only a handful of critical changes that occurred—five, six or seven genes—that allowed a weedy species to become desirable, such as changes in fruit size or corn ear number, those sorts of things.
    A. a person's willingness to listen and pay attention to something
    B. the organ of hearing and balance in humans
    C. a head of maize in North America
    D. the seed-bearing head or spike of a cereal plant

Unit 5
GM Technology

● **Section Two**

Directions: There are two or three meanings for each semi-technical term underlined in the following sentences. Choose the correct one according to the context.

1. On a foggy morning last fall, at a continuing education seminar on mushrooms, a biologist named Yinong Yang took the podium to deliver news of a possible solution for the browning (becoming brown by genes churning out enzyme in the fruiting body of certain plants; the process of making something brown by cooking or burning) problem.

2. They understood the enormous commercial implications when Yang showed them photos comparing brown, decayed mushrooms with pristine white, CRISPR-engineered A. bisporus, the all-purpose strain (a force tending to pull or stretch something to an extreme or damaging degree; a breed, stock, or variety of a plant developed by breeding; a variety of a particular abstract thing) that annually accounts for more than 900 million pounds of white button, cremini and Portobello mushrooms.

3. CRISPR allows you to knock out (destroy a machine or damage it so that it stops working; knock down a boxer for a count of ten; eliminate) any gene or, with a little more effort, to add a desirable trait by inserting a gene in a specific place in a genome.

4. The fact that we got regulatory clearance (the removal of buildings, people, or trees from land so as to free it for alternative uses; the discharge of a debt; official authorization for something to proceed or take place) and can go almost immediately from the greenhouse to the field is a big plus.

5. Researchers at the small Minnesota-based biotech firm Recombinetics have genetically blocked the biological signal that governs (conducts the policy, actions, and affairs of a state; controls, influences, or regulates an action or course of events; serves to decide a legal case) the growth of horns in Holstein cows.

6. The USDA has given regulatory clearance to plant a potato variety (the quality of being different or diverse; a taxonomic category that ranks below species, its members differing from others of the same species in minor but permanent or heritable characteristics) and two soybean varieties, so the potato and one of the soybean varieties are in the field this year.

7. The cutting is then accomplished by the Cas9 enzyme, originally isolated from bacterial cultures (the arts and other manifestations of human intellectual achievement regarded collectively; preparation of cells obtained in an artificial medium containing nutrients;

cultivation of plants) in yogurt, which rides in on the back of the guide RNA.

8. If you make a piece of guide RNA that is 20 letters long, it will find its mirror <u>sequence</u> (the order in which amino-acid or nucleotide residues are arranged in a protein, DNA, etc.; an infinite ordered series of numerical quantities; a part of a film dealing with one particular event or topic) of DNA amid the string of 30 million letters that spell out the *Agaricus* mushroom genome.

9. Any time DNA from two different individuals comes together during <u>reproduction</u> (the action of making a copy of something; the production of offspring by a sexual or asexual process), the DNA gets scrambled in a process known as chromosomal reassortment.

10. Rather than <u>crossing</u> (placing something crosswise; going to the other side of; causing to interbreed with one of another species, breed, or variety) the wild varieties with the domesticated strains, which would require a 10-year breeding regime, he says, "maybe we can just go in and treat those genes and domesticate the wild variety."

● **Section Three**

Directions: Match the Chinese terms with their English equivalents.

1. 转基因生物　　　　　　A. gene-editing tool
2. 单一不饱和脂肪　　　　B. browning gene
3. 菜籽油　　　　　　　　C. human germ-line modification
4. 化学诱变剂　　　　　　D. gene-edited organism (GEO)
5. 基因编辑工具　　　　　E. genetically modified organism (GMO)
6. DNA双螺旋结构　　　　F. monosaturated fat
7. 褐变基因　　　　　　　G. canola oil
8. 堆肥　　　　　　　　　H. chemical mutagen
9. 人类生殖细胞修改　　　I. base pair
10. 卵细胞　　　　　　　　J. nucleic acid
11. 基因编辑生物　　　　　K. double helix of DNA
12. 原理验证　　　　　　　L. mushroom house
13. 核酸　　　　　　　　　M. compost
14. 碱基对　　　　　　　　N. egg cell
15. 蘑菇栽培室　　　　　　O. proof of principle

III. Questions for Discussion

Directions: Work in groups and discuss the following questions.

1. Are the general public entitled to being informed of the truth of transgenic crop plants?
2. How do you see the relationship between researchers of GM technology and agricultural companies?
3. Which do you prefer, GEO food or conventionally grown food? Why?

Part II Extensive Reading

Text B

The American Chestnut's Genetic Rebirth

A foreign fungus nearly wiped out North America's once vast chestnut forests. Genetic engineering can revive them.

By William Powell

In 1876, Samuel P. Parsons received a shipment of chestnut seeds from Japan and decided to grow and sell the trees to orchards. Unbeknownst to him, his shipment likely harbored a stowaway that caused one of the greatest ecological disaster ever to befall eastern North America. The trees probably concealed spores of a pathogenic fungus, *Cryphonectria parasitica*, to which Asian chestnut trees—but not their American cousins—had evolved resistance. *C parasitica* effectively strangles a susceptible tree to death by forming cankers—sunken areas of dead plant tissue—in its bark that encircle the trunk and cut off the flow of water and nutrients between the roots and leaves. Within 50 years this one fungus killed more than three billion American chestnut trees.

Before the early 1900s the American chestnut constituted about 25 percent of hardwood trees within its range in the eastern deciduous forests of the U.S. and a sliver of Canada—deciduous forests being those composed mostly of trees that shed their leaves in the autumn. Today only a handful of fully grown chestnuts remain, along with millions of root stumps.

In its prime, the American chestnut was a keystone species, crucial to the health of a multitude of organisms in its ecosystem. Many different birds, insects and small mammals nested in its branches and burrowed into its bark. Bears, deer, turkeys, blue jays, squirrels and other animals ate the large, nutritious chestnuts. After losing so many mature chestnut trees, wildlife populations declined and became less diverse. The oaks that have since replaced the chestnut cannot support as many animals; the acorns they produce are only half as nutritious. And chestnuts once generated larger quantities of nuts than oaks do today, in part because they flowered after frosts that might have destroyed delicate buds.

The American chestnut also had great economic value. Its nuts can be used for food or ethanol fuel. Because the American chestnut grows quickly, has sturdy, straight-grained wood and is very rot-resistant, it provides excellent timber. In fact, if the chestnut were still abundant, most decks would likely be made from its wood instead of from pressure-treated lumber, which often contains heavy metals and other preservatives that endanger the environment and people's health when they find their way into soil and food. Last, the American chestnut has been an especially beloved tree, immortalized in poetry, songs, books, street signs, and the names of many schools, hotels and parks across the country.

We do not have to stand by as the American chestnut becomes a distant memory for most people. The culmination of decades of research suggests that science can restore the tree and all the resources it once offered people and wildlife. After a century of ineffective efforts to combat chestnut blight, two approaches are now meeting with some success. One strategy attempts to create blight-resistant American chestnuts with an ancient horticultural technique: hybridization. By mating American chestnuts with far smaller, fungus-resistant Chinese chestnuts, researchers "backcross" the resulting hybrids with other American chestnuts to Americanize the trees as much as possible while, it is hoped, keeping all the genes responsible for blight resistance. In addition to being rather imprecise, however, backcross breeding requires many generations and thousands of trees to produce individuals suitable for restoration.

For those reasons, my many collaborators and I are focusing on a second approach, which relies on altering the chestnut tree's DNA in a much more exact way than traditional breeding and which has the potential to produce more fungus-resistant trees more quickly. By borrowing genes from wheat and the Chinese chestnut, among other plants, and inserting them into the American chestnut's genome, we have created hundreds of transgenic trees, some of which defend themselves against *C. parasitica* as well as, if not better than, their Asian counterparts.

Compared with other efforts to revive endangered or extinct species with genetic engineering and related biotechnologies, the efforts to reinstate the American chestnut face far fewer hurdles and offer much clearer benefits. Unlike cloned mammoths and pigeons, trees do not require surrogate mothers, parenting or socialization. And as a massive organism that is home to many others, the American chestnut can improve the health of the forest more than any one animal.

- **Seeds of Salvation**

In 1983, when I became a graduate student working with plant pathologist Neal Van Alfen, then at Utah State University, I began to develop a deep appreciation and sympathy for the magnificent chestnut tree and its demise at the hands of an exotic pathogen.

In 1989, when I had moved to the S.U.N.Y. College of Environmental Science and Forestry, Stan Wirsig of the American Chestnut Foundation approached my colleague Charles Maynard and me with a proposition. He wanted to complement the foundation's ongoing chestnut tree hybridization program with a new restoration project focused on genetic engineering, which was a cutting-edge technology at the time and promised a speedier and more precise way to create resistant American chestnuts. One of my tasks was to find a gene that could endow the trees with resistance to *C. parasitica* while Maynard and Scott Merkle of the University of Georgia developed the techniques that would allow us to introduce that gene to chestnut tree embryos—tiny bundles of swiftly multiplying cells that would eventually grow into adult trees. If everything worked as planned, the young trees would grow into sturdy adults with the ability to battle the fungus.

At that time, no one had ever tried to genetically engineer a tree to fight a virulent fungus, but we had a few clues about how to get started. Over the years researchers had learned some important details about how *C. parasitica* damages chestnut trees. The pathogen grows feathery lattices of fungal tissue called mycelial fans that produce oxalic acid, which eats through the tree's bark to make room for the fungal invasion. As the fungus wedges its way into the tree, a canker girdles the trunk.

Initially we focused on finding a way to weaken the mycelial fans. We knew that the immune systems of many plants and animals contain small chains of amino acids known as antimicrobial peptides (AMPs) that can disrupt fungal cells. Using AMP genes in the African clawed frog as a model, we assembled genes from scratch to produce AMP peptides that could fight *C. parasitica*. We hoped that if we could engineer the chestnut trees to produce even small amounts of these AMPs, they would make mycelial fans go slack and thereby render them benign. Such peptides are notoriously unstable molecules, though, so we needed a backup plan.

Around the same time, a then graduate student named Kim Cameron stopped by my office and dropped off a book summarizing many of the studies presented at the recent annual meeting of the American Society of Plant Biologists. When I read about a study conducted by Ousama Zaghmout and Randy Allen, I had a eureka moment❶. The study described a wheat gene for an enzyme called oxalate oxidase (OxO), which breaks down oxalic acid—the very same caustic substance produced by the chestnut blight fungus. Even better, the researchers had worked out a way to introduce this gene into other plants. They put the gene into *Agrobacterium*, a microbe that can inject DNA into the command center of plant cells, and exposed plants to clones of that microbe. The resulting transgenic plants became resistant to an acid-spewing fungus known as *Sclerotinia sclerotorium*. Maybe we could do something similar with the American chestnut.

We could not test either approach on chestnuts at that point, because we were still figuring out how to grow the finicky chestnut in the laboratory. So we decided to achieve a proof of principle in a different tree—the hybrid poplar, which was well studied and often used in experiments. Haiying Liang, then a graduate student at the College of Environmental Science and Forestry, would deliver both the *OXO* gene and our AMP gene, and when the trees were old enough, we would infect them with *Septoria musiva*, a fungus that produces a good deal of oxalic acid and can cause leaf spot and canker diseases in hybrid poplars. Most of the trees treated in this way remained relatively healthy. We had made one tree fungus-resistant with genetic engineering. Now we needed to do it with the right tree and the right fungus.

---

❶ Eureka, from Greek, means "I have found it". It is said to have been uttered by Archimedes when he hit upon a method of determining the purity of gold, jumped out of his tub, and ran naked through the streets, shouting to his startled neighbors: "Eureka!" Now it refers to a cry of joy or satisfaction when one finds or discovers something, or solves a problem.

While Liang was conducting the poplar experiments, Linda McGuigan, also then a graduate student at the college, set to work figuring out how to raise chestnut trees from embryos in the lab. Some plants, like carrots and petunias, are remarkably easy to grow in the lab. The American chestnut was not one of these cooperative plants. McGuigan spent two and a half years learning how to successfully introduce the wheat gene into chestnut embryos using Agrobacterium and to subsequently shepherd the embryos into young adulthood in the lab. He learned how to control lighting, humidity and temperature to mimic what would normally happen inside a chestnut seed and fine-tuned the delivery of various hormone cocktails at different stages of the miniature tree's early development to induce growth of roots and shoots.

In 2006 we were able to plant the first transgenic American chestnut trees in experimental fields sectioned off from the forest. It takes at least two to three years for the trees to reach a size at which we can challenge them with the blight fungus. We had attached the *OXO* gene to a promoter—a kind of genetic switch that controls how often a cell reads the instructions in a gene—to limit the production of OxO to certain tissues. We were hoping the resulting low levels of the enzyme would be sufficient to take on the fungus without causing any unwanted side effects. Unfortunately, we were mistaken. This first line of trees was not able to resist the fungus; they died a little slower than is typical but ultimately succumbed to their illness.

By 2012 we had designed a new promoter for the *OXO* gene and engineered a new line of trees that produced much more of the acid-degrading enzyme. Success! These trees evaded disease almost as well as the Chinese chestnut, which had evolved resistance on its own. We have now developed a way of gauging disease resistance by testing the leaves of chestnut trees that are only a few months old, so we no longer have to wait three years to see if our experiments are working. In this test, we make small cuts in leaves, infect them with fungus and wait for a circle of decaying tissue to spread from the wound. The smaller the spot of death, the more resistant the tree. Some of our newest trees appear to be even more resistant than the Chinese chestnut. We need to confirm this finding as the trees get older, but it appears that the gene we borrowed from wheat has exceeded our expectations.

Unit 5
GM Technology

People often ask us why we do not simply find the genes that make the Chinese chestnut resistant and use them instead of the wheat gene. When we first started our research, no one had thoroughly studied the Chinese chestnut genome, and it would have taken too much time and too many resources to locate the numerous different genes responsible for a complex trait like blight resistance. Each of those genes would contribute only a small portion of the tree's ability to battle the fungus, and any one of them would probably have been ineffective as a defense on its own.

At this point, however, scientists have identified 27 genes that might be involved in the Chinese chestnut's blight resistance. So far two of these genes each appear to endow trees with an intermediate level of resistance. Testing is ongoing with the other candidate genes. Eventually we hope to fortify American chestnuts with many different genes that confer resistance in distinct ways. Then, even if the fungus evolves new weapons against one of the engineered defenses, the trees will not be helpless.

● Going Out on a Limb❶

Today more than 1,000 transgenic chestnut trees are growing in field sites, mostly located in New York State. The next hurdle for American chestnut restoration involves the federal regulatory process. Before we can plant trees in the forest, the FDA, USDA and EPA will want to make sure that genetically engineered chestnut trees are not significantly different from typical trees in some unexpected way. As opposed to hybridized trees, which are genetically quite different from American chestnuts because they have large chunks of Chinese chestnut DNA, our transgenic trees have only a few new genes. Preliminary tests show that the roots of typical chestnut trees and engineered trees form the same kinds of symbiotic relations with helpful fungi and that similar communities of smaller plants grow underneath the canopies of both modified and unmodified trees. Likewise, the same insect species visit both transgenic and typical chestnut trees, and nuts from both types of trees have the same nutritional composition. Once such tests are complete, we will petition the FDA, USDA and EPA for the same unregulated status that they give to genetically

---

❶ Out on a limb means in or into a position where one is not joined or supported by anyone else.

engineered crops.

A final hurdle is public acceptance. Encouragingly, many people who are typically opposed to genetic modification make an exception for the American chestnut tree. Some people reason that because humans caused the demise of the chestnut in the first place, humans should fix it. Others are accepting because we are not seeking profit and are not patenting the trees.

Many people are also happy to learn that the environmental risks of American chestnut restoration are negligible. The chances of transgenic chestnut tree pollen spreading introduced genes to other plant species are very small. Ideally, some of the transgenic pollen will spread resistance to at least a fraction of the remaining American chestnut stumps that manage to flower, rescuing as much of their total genetic diversity as possible. If the stumps do benefit, they could spawn a blight-resistant population that, over the centuries, could return this once towering keystone species to its former glory in the eastern forests.

Chestnut blight is not the only enemy of biodiversity that genetic engineering can eradicate. We are losing the battle against many other exotic pests such as the hemlock woolly adelgid—a bug that sucks the sap from hemlock trees—and the emerald ash borer—a metallic green beetle whose larvae tunnel under the bark of ash trees—as well as the pathogens responsible for sudden oak death and walnut thousand cankers disease, to name a few. To turn the tables❶, we have to act quickly, and in most cases, traditional breeding techniques are just too slow to make a difference. Now, more than ever, we need genetic engineering in our toolbox to maintain diverse and healthy forests.

Completely restoring the American chestnut to its previous status as a king of the forest is a centuries-long endeavor. Once the chestnut trees pass regulatory and public approval, a good place to begin restoration is on reclamation lands. With the help of the Forest Health Initiative and Duke Energy, test plots are now being planted on mine reclamation sites. Other areas might include abandoned farmland and historic locations that once had abundant chestnut trees. And perhaps some individuals will

---

❶ Turn the tables means reversing one's position relative to someone else, especially by turning a position of disadvantage into one of advantage.

want to have these iconic trees in their own yards. An old Chinese proverb says, "One generation plants a tree, the next generation enjoys its shade." In the case of the American chestnut, we are that first generation.

(Excerpt from *Scientific American*, March 2014)

Exercises

I. Translate the following technical terms into English.

1. 阔叶林 _____
2. 落叶林 _____
3. 直纹木 _____
4. 耐腐蚀的 _____
5. 压力处理的木材 _____
6. 防腐剂 _____
7. 枯萎病抗性 _____
8. 杂交树 _____
9. 回交育种 _____
10. 抗菌性 _____
11. 转基因树 _____
12. 代孕母亲 _____
13. 植物病理学家 _____
14. 外来病原体 _____
15. 胚芽 _____
16. 氨基酸 _____
17. 腐蚀性物质 _____
18. 酸降解酶 _____
19. 共生关系 _____
20. 花粉 _____

II. Translate the following paragraphs into Chinese.

We do not have to stand by as the American chestnut becomes a distant memory for most people. The culmination of decades of research suggests that science can restore the tree and all the resources it once offered people and wildlife. After a century of ineffective efforts to combat chestnut blight, two approaches are now meeting with some success. One strategy attempts to create blight-resistant American chestnuts with an ancient horticultural technique: hybridization. By mating American chestnuts with far smaller, fungus-resistant Chinese chestnuts, researchers "backcross" the resulting hybrids with other American chestnuts to Americanize the trees as much as possible while, it is hoped, keeping all the genes responsible for blight resistance. In addition to being rather imprecise, however, backcross breeding requires many generations and thousands of trees to produce individuals suitable for restoration.

For those reasons, my many collaborators and I are focusing on a second approach, which relies on altering the chestnut tree's DNA in a much more exact way than traditional breeding and which has the potential to produce more fungus-resistant trees more quickly. By borrowing genes from wheat and the Chinese chestnut, among other plants, and inserting them into the American chestnut's genome, we have created hundreds of transgenic trees, some of which defend themselves against *C. parasitica* as well as, if not better than, their Asian counterparts. If the U.S. Department of Agriculture, the Environmental Protection Agency, and the Food and Drug Administration approve our trees—which could happen as soon as five years from now—they will be the very first transgenic organisms used to restore a keystone species to its native environment.

Text C

Saltwater Solution

Farmland is being ruined by salty water. Rice and fruits, genetically modified to survive salt, could feed millions.

By Mark Harris

Eric Rey pulls a plastic container half full of cooked rice out of his briefcase. They smell like normal rice. When I gingerly raise a few grains to my lips, they even taste like normal rice: soft, chewy and a little bland. I have to stop myself from reaching for a bottle of soy sauce here in the kitchen of Arcadia Biosciences' offices in Seattle—Rey is the chief executive of the biotechnology company—to add a little salt.

My desire for extra flavor is a bit odd because this rice was grown in a salty brine that would kill most plants on the earth. The rice plants were genetically engineered to survive the chemical, mimicking unusual plants called halophytes that flourish on ocean bays, inlets and marshy shorelines. I'm surprised the grains in my mouth don't make my tongue curl. I try a blind taste test comparing them with unmodified rice grown in freshwater, and I can't tell the difference.

"Rice is the most valuable crop in the world," measured by the amount produced in 2012, Rey says, but "in parts of China where the salinity has gone up and up, they basically can't grow crops anymore." Rey believes that new understanding of the genes that help halophytes cope with huge doses of salt, combined with modern biotech methods of inserting those genes into rice and other plants, could hold a key to feeding our planet's growing population.

Nearly a quarter of the world's irrigated areas suffer from salty soil caused by poor irrigation practices. Sea-level rises also threaten tens of millions of hectares more farmland with saltwater intrusion. If healthy crops could be grown in such salty regions, they might provide food for tens of millions of people, a vital step toward supporting the extra two billion mouths expected on the earth by midcentury.

This is no pipe dream❶, says Eduardo Blumwald, a plant biologist at the University of California, Davis, whose work forms the basis of Arcadia's rice. "I believe it's now feasible to grow crops in low-quality, brackish and recycled water, even diluted seawater," he says. He and a few other scientists around the world are transferring genes from naturally salt-tolerant halophytes into everyday crops—not just rice but also wheat, barley and tomatoes.

For these seeds of salvation to take root, however, they will have to move out of greenhouses and prove they can thrive amid real-world storms, droughts and predatory insects. They will also need to survive a tempest of safety and regulatory questions from politicians, scientists and farmers. Even if the plants themselves are delicious, genetic engineering can leave a nasty taste in people's mouths. They worry the genes may be transferred to other organisms, with unforeseen effects. Furthermore, points out Janet Cotter, an environmental consultant, creating food that can be grown in salty conditions simply encourages more poor irrigation practices.

---

❶ A pipe dream is a fanciful hope or plan that you know will never really happen.

• A Salty Tale

Halophytes, whose very name means "salt plants", can survive in water ranging in salinity from a stiff Bloody Mary❶ to full-on seawater. Mangroves are halophytes. Early attempts to popularize halophytes tried to stimulate a market by touting mangroves as a building material, oil-rich succulent halophytes for biofuels or salt-tolerant bushes for animal forage. In 1998 researchers wrote an article in *Scientific American* envisioning large-scale halophyte farms around the world to feed people. But in the absence of any developed markets for the niche crops they offered, such farms were doomed to failure.

By the time Blumwald began work on halophytes in the mid-1990s, they had been largely dismissed as botanical curiosities. "Most agricultural scientists never thought about salinity," he says. "They were thinking about making food bigger, rounder, more colorful, sweeter."

Blumwald, however, became interested in a protein found in these plants that is called an antiporter. It accelerates the exchange of sodium (salt) and hydrogen ions across a plant's cell membranes. When sodium in water is absorbed by the plant, it disrupts enzymes, the transport of water around the plant and, ultimately, photosynthesis itself. Blumwald found that by genetically engineering everyday species to produce large amounts of this antiporter, he was able to breed plants that could grow in water a third as salty as seawater, with few ill effects. The antiporter pushed sodium ions into vacuoles, sealed-off spaces within cells, where they could do no harm. In some natural halophytes, these vacuoles become so big that they are called salt bladders.

When Blumwald boosted antiporter levels in some English heirloom tomatoes, the plants grew in water that was "four times as salty as chicken soup", he says. And they produced red, round, sweet, juicy fruit, each weighing several ounces. But while Blumwald's creations thrived in the laboratory, they struggled in the real world. "Everything works in the greenhouse, where you have a relative humidity of 40 percent or more," Blumwald says. As humidity decreases, however, plants lose more

---

❶ Stiff here means (of an alcoholic drink) strong. Bloody Mary is the nickname of Mary I of England, and now it is often used to refer to a drink consisting of vodka and tomato juice.

moisture from their leaves and defensively close pores. So growing plants is much harder, he notes, "When you go to the field, with a humidity of 5 percent and much less water."

The problem is that an ability to shed salt is not the only requirement for growing well in salty soil. Plants possess thousands of genes, involved in many biological processes, that can help the organism cope with many kinds of stress, such as heat, drought or salinity. To grow in salty conditions, a plant needs to have multiple genes that change their activity to protect the plant when growing conditions become challenging. There is no single magic bullet, says Simon Barak, a senior lecturer in plant sciences at Ben-Gurion University of the Negev in Israel, "but we have developed a computational method to sift through those genes and see which are most likely to be involved in stress tolerance."

Barak constructed a stress gene database, gathering data from published experiments on the plant Arabidopsis thaliana. Using statistical analyses that allowed him to rank the importance of each gene for plant survival under conditions such as high heat, he identified a number of promising candidate genes. Then Barak's group ran lab tests on plants with mutated versions of those genes to see how the vegetation coped with harsh conditions. Mutants that showed tolerance to drought, salt or heat were then targeted for further study.

Other researchers have also homed in on salt survival by blending biology with statistics and computer science. A few years ago, for example, while working at the Central Salt & Marine Chemicals Research Institute in Gujarat, India, geneticist Narendra Singh Yadav found a number of genes associated with salt tolerance in another halophyte, salicornia. He did not know exactly what the genes did, only that his analysis suggested they played an important role. To test his theory, Yadav inserted two of these genes into tobacco, a plant usually quite vulnerable to salt. When grown in water about a third as salty as seawater, the transgenic plants germinated better, had longer roots and shoots, and were larger and leafier than unmodified plants. Although they did not develop visible salt bladders, the plants had lower levels of harmful molecules called reactive oxygen species that accumulate under salt stress. Yadav is now based in Israel with Barak, and his former research group is working on a salt-tolerant version of cotton in Gujarat. "And I think there are still a lot more genes to discover," he says.

The important thing, Blumwald says, is "to be intelligent without being stupidly optimistic". His group at U.C. Davis has a dozen greenhouses running experiments on thousands of different transgenic plants, from alfalfa and pearl millet to peanuts and rice. Most are modifications of successful commercial crops, and each experiment tries to replicate natural, stressful conditions. Massive fans simulate erratic winds, water is delivered at irregular intervals or in pulses like storms, and salt and heat are applied.

● **Natural Fears**

Genetic Modification (GM), however, remains controversial in many parts of the world. Cotter prefers a breeding system called marker-assisted selection that uses genomic tools to identify genes for salt tolerance in wild versions of crop plants, then naturally breeds those plants with domesticated ones to reintroduce the gene into plants grown on farms. Timothy Russell, who is an agronomist working in Bangladesh for the International Rice Research Institute, is also skeptical. "There's not a huge problem with GM in my mind, but it is a lot easier to get a conventionally bred variety into the market," he says. "We think that we can get reasonably good tolerance to salinity using conventional techniques. Why go down a more complicated way when it's not really necessary?"

One good reason to use GM, advocates say, is that it is faster. Breeding, selecting and rebreeding take time. Genetically engineered salt-tolerant crops will likely beat conventionally bred plants to market, probably within the next four years. The salt-tolerant rice I tasted from Arcadia Biosciences is already halfway through its final field trials in India and is headed for regulatory approval there. The plant produces 40 percent more grain than today's rice in water a tenth as salty as the sea, and Rey expects a subsequent strain to be twice as tolerant again.

It's a small start, Blumwald feels: "It's a step in the right direction. Feeding billions more people in the future will require not one success like this but dozens or hundreds."

(Excerpt from *Scientific American*, July 2016)

Unit 5
GM Technology

∽ Exercises ∾

▮ Reading Comprehension

Directions: Answer the following questions based on the information from the text.

1. Tell the various attitudes to rice and fruits that are genetically modified in order to survive salt.
2. How does antiporter work in salty soil?
3. Why do researchers say that an ability to shed salt is not the only requirement for plants to grow well in salty soil?
4. What are the findings reported by Blumwald, Barak, and Yadav?
5. In what ways do the greenhouses simulate the natural conditions for plants to grow?

Unit 6
Behavioral Science

导读

本单元的文章均涉及行为科学研究领域的热门话题。Text A详细讲解了我们的行为习惯是如何形成的，并通过实验验证了习惯的形成机制。Text B通过大量的实验阐述了婴儿的天生语言学家本质，认为婴儿学习并不是一个被动的过程，社会互动才是掌握语言的基本条件。Text C详细阐述了自控能力对于一个人成功的重要性。

Part I Intensive Reading

Text A

Good Habits, Bad Habits

Researchers are pinpointing the brain circuits that can help us form good habits and break bad ones.

By Ann M. Graybiel and Kyle S. Smith

Every day we all engage in a surprising number of habitual behaviors. Many of them, from brushing our teeth to driving a familiar route, simply allow us to do certain things on autopilot so that our brains are not overtaxed by concentrating on each brushstroke and countless tiny adjustments of the steering wheel. Other habits, such as jogging, may help keep us healthy. Regularly popping treats from the candy dish may not. And habits that wander into the territory of compulsions or addictions, such as overeating or smoking, can threaten our existence.

Even though habits are a big part of our lives, scientists have had a hard time pinning down how the brain converts a new behavior into a routine. Without that knowledge, specialists have had difficulty helping people break bad habits, whether with medicines or other therapies.

New techniques are finally allowing neuroscientists to decipher the neural mechanisms that underlie our rituals, including defining our so-called habit circuits—the brain regions and connections responsible for creating and maintaining our routines. The research suggests that by deliberately conditioning our brain, we might be able to control habits, good and bad. That promise springs from one of several surprises: That even when it seems we are acting automatically, part of our brain is dutifully monitoring our behavior.

• What Is a Habit, Really?

Habits seem to stand out as clear-cut actions, but neurologically, they fall along a continuum of human behavior. At one end of that continuum are behaviors that can be done automatically enough to let us free up brain space for different pursuits. Others can command a lot of our time and energy. Our habits emerge naturally as we explore our physical and social environments and our inner feelings.

We all begin this process when we are very young. Yet it comes with a trade-off that can work against us. The more routine a behavior becomes, the less we are aware of it. Witness computer gaming, Internet gambling, and constant texting and tweeting—and of course alcohol and drug use. A repetitive, addiction-driven pattern of behavior can take over part of what had been deliberate choice. Neuroscientists are still grappling with whether addictions are like normal habits, only more so, although they certainly can be thought of as extreme examples at the other end of the continuum, so can certain neuropsychiatric conditions such as obsessive-compulsive disorder—in which thoughts or actions become all-consuming—and some forms of depression, in which negative thoughts may run in a continuous loop. And extreme forms of habit may be involved in autism and schizophrenia, in which repetitive, overly focused behaviors are a problem.

• Deliberate Behavior Becomes Routine

Although habits fall along different parts of the behavior spectrum, they share certain core features. Tell yourself to "stop doing that", and most of the time the lecture fails! Part of the reason may be that this critique usually happens too late, after the behavior plays out and its consequences are being felt.

This stubbornness, in particular, has been a clue to uncovering the brain circuitry responsible for habit formation and maintenance. Habits become so ingrained that we perform them even when we do not want to, in part because of what are called "reinforcement contingencies❶". Say you do A, and then you are rewarded somehow. But if you do B, then you are not rewarded or are even punished. These consequences of our actions—the contingencies—push our future behavior one way or another.

Signals discovered in the brain seem to correspond to this reinforcement-related learning, as shown in studies originally conducted by Wolfram Schultz and Ranulfo Romo, both then at the University of Fribourg in Switzerland, and today modeled by computational scientists. Particularly important are "reward-prediction error signals", which, after the fact, indicate the mind's assessment of how accurate a prediction about a future reinforcement actually turned out to be. Somehow the brain computes these evaluations, which sculpt our expectations and add or subtract value from particular courses of action. By monitoring our actions internally and adding a positive or negative weight to them, the brain reinforces specific behaviors, shifting actions from deliberate to habitual—even when we know we should not gamble or overeat.

In the Graybiel lab at the Massachusetts Institute of Technology, our group began experiments to decipher which brain pathways were involved and how their activity might change as habits formed. First, we needed an experimental test for determining whether a behavior is a habit. British psychologist Anthony Dickinson had devised one in the 1980s that is still widely used. He and his colleagues taught lab rats in a test box to press a lever to receive a food treat as a reward.

When the animals had learned this task well and were back in their cages, the experimenters "devalued" the reward, either by letting the rats eat the reward to the point of oversatiation or by giving them a drug that produced mild nausea after

---

❶ Reinforcement contingency is a positive or negative relationship between the reinforcer and the response.

the reward was eaten. Later on, they brought the rats back to the experimental box and gave them the choice of pressing the lever or not. If a rat pressed the lever even though the reward was now sickening, Dickinson considered the behavior to be a habit. But if a rat was "mindful"—if we can speak of mindfulness in a rat—then it did not press the lever, as though it realized that the reward was now unpleasant; it had not formed a habit. The test gave scientists a way to monitor whether or not a shift from purposeful to habitual behavior had occurred.

● **Imprinting a Habit on the Brain**

By using variations of this basic test, researchers, including Bernard Balleine of the University of Sydney and Simon Killcross of the University of New South Wales in Australia, have found clues suggesting that different brain circuits take the lead as deliberate actions become habitual. New evidence from experiments on rats, as well as on humans and monkeys, now points to multiple circuits that interconnect the neocortex—regarded as the crowning glory of our mammalian brain—and the striatum, at the center of the more primitive basal ganglia, which sits at the core of our brain. These circuits become more or less engaged as we act deliberately or habitually.

We taught rats and mice to perform simple behaviors. In one task, they learned to run down a T-shaped maze once they heard a click. Depending on an audio "instruction" cue that then sounded as they ran, they would turn left or right toward the top of the T and run to that end to receive one kind of reward or another. Our goal was to understand how the brain judges the pros and cons of behaving in a particular way and then stamps a sequence of behavior as a "keeper"—a habit. Our rats certainly did develop habits! Even when a reward had become distasteful, the rats would run to it when the instruction tone sounded.

To figure out how the brain stamps a behavior as one to make a habit, the M.I.T. lab began recording the electrical activity of small collections of neurons (brain cells) in the striatum. What our group found surprised us. When the rats were first learning the maze, neurons in the motor-control part of the striatum were active the whole time the rats were running. But as their behavior became more habitual, neuronal activity began to pile up at the beginning and end of the runs and quieted down during most of the time in between. It was as though the entire behavior had become packaged, with the striatal cells noting the beginning and end of each run. This was an unusual pattern; what seemed to be happening was that the striatal cells were malleable and could help package movements together while leaving relatively few "expert cells" to handle the details of the behavior.

This pattern reminded us of the way the brain lays down memories. We all know how helpful it is to remember a string of numbers as larger units instead of one by one—such as thinking of a phone number as "555-1212" instead of "5-5-5-1-2-1-2". The late American psychologist George A. Miller coined the term "chunking" to refer to this packaging of items into a memory unit. The neural activity we observed at the beginning and end of a run seemed similar. It is as though the striatum sets up boundary markers for chunks of behavior—habits—that the internal evaluation process has decided should be stored. If true, this maneuver would mean that the striatum essentially helps us combine a sequence of actions into a single unit. You see the candy dish, and you automatically reach for it, take a treat and eat it "without thinking".

Researchers have also identified a "deliberation circuit", which involves another part of the striatum and is active when choices are not made on autopilot and instead require some decision making.

To understand the interplay between these deliberation and habit circuits, our group's Catherine Thorn recorded signals in both circuits simultaneously. As the animals learned a task, activity in the deliberation part of the striatum became strong during the middle of the runs, especially when the rats had to decide which way to turn at the top of the T, based on the instruction tone. This pattern was almost the exact opposite of the chunking pattern that we had seen in the habit striatum. And yet the activity did recede as the behavior became fully habitual. The pattern means that as we learn habits—at least as rats do—habit-related circuits gain strength, but changes in related circuits occur, too.

Because the striatum works together with a habit-related part of the neocortex at the front of the brain known as the infralimbic cortex, we then recorded activity in that region. This was an eye-opener as well. Even though we saw the beginning-and-end pile-up of activity in the habit striatum, during the initial learning period we saw very little change in the infralimbic cortex. It was not until the animals had been trained for a long time and the habit became fixed that the infralimbic activity changed. Strikingly, when it did, a chunking pattern then developed there, too. It was as though the infralimbic cortex was the wise one, waiting until the striatal evaluation system had fully decided that the behavior was a keeper before committing the larger brain to it.

- **Stop That!**

We decided to test whether the infralimbic cortex has online control over whether a habit can be expressed by using a new technique called optogenetics. With this technique, we could place light-sensitive molecules in a tiny region of the brain, and then, by shining light on the region, we could turn the neurons in that region on or off. We experimented with turning off the infralimbic cortex in rats that had fully acquired the maze habit and had formed the chunking pattern. When we turned off the neocortex just for a few seconds while the rats were running, we totally blocked the habit.

The habit could be blocked rapidly, sometimes immediately, and the habit blockade endured even after the light was turned off. The rats did not stop running in the maze, however. It was just the habitual runs to the devalued reward that were gone. The animals still ran just fine to reach the good reward on the other side of the maze. In fact, as we repeated the test, the rats developed a new habit: running to the good-reward side of the maze no matter what cue they were given.

When we then inhibited the same tiny piece of infralimbic cortex, we blocked the new habit—and the old habit instantly reappeared. This return of the old habit happened in a matter of seconds and lasted for as many runs as we tested, without our having to turn off the infralimbic cortex again.

Many people know the feeling of having worked hard to break a habit only to have it come back, full-blown, after a stressful time or after one relapse. When Russian scientist Ivan Pavlov studied this phenomenon in dogs many years ago, he concluded that animals never forget deeply conditioned behaviors such as habits. The most they can do is suppress them. We are finding the same stubbornness of habits in our rats. Yet remarkably, we can toggle the habits on and off by manipulating a tiny part of the neocortex during the actual behavior. We do not know how far this control could reach. For example, if we taught the rats three different habits in a row, then blocked the third one, would the second habit appear? And if we then blocked the second one, would the first one appear?

A key question was whether we could prevent a habit from forming in the first place. We trained rats just enough to have them reach the correct end of the T but not enough for the behavior to settle in as a habit. We then continued the training, but during each run we used optogenetics to inhibit the infralimbic cortex. The rats continued running well in the maze, but they never acquired the habit, despite many days of overtraining that usually would have made the habit permanent. A group of control rats that underwent the same training without the optogenetic interruption did form the habits normally.

- **Breaking Bad Habits**

Our experiments offer some curious lessons. First, no wonder habits can be so difficult to break—they become laid down and marked as seemingly standardized chunks of neural activity, a process involving the work of multiple brain circuits.

Yet surprisingly, even though habits seem nearly automatic, they are actually under continual control by at least one part of the neocortex, and this region has to be online for the habit to be enacted. It is as though the habits are there, ready to be reeled off, if the neocortex determines that the circumstances are right. Even if we are not conscious of monitoring our habitual behaviors, we have circuits that actively keep track of them on a moment-to-moment basis. We may reach out for the candy dish without "thinking", but a surveillance system in the brain is at work, like a flight-monitoring system in an airliner.

So how close are we to helping people clinically? It will likely be a long time before anyone can flip a switch to zap away our pesky habits. The experimental methods that we and others are using cannot yet be translated directly to people.

Unit 6
Behavioral Science

But neuroscience is changing at lightning speed, and those of us in the field are closing in on something truly important: the rules that habits work by. If we can fully understand how habits are made and broken, we can better understand our idiosyncratic behaviors and how to train them. It is also possible that our expanding knowledge could even help people at the severe end of the habit spectrum, providing clues for how to treat obsessive-compulsive disorder, Tourette's syndrome[1], fear or post-traumatic stress disorder.

Drug treatments and other emerging therapies could possibly do the trick to help with such harmful habits. But we are also impressed by how the lessons we have learned from this brain research support behavioral therapy strategies, which are often suggested for helping us to establish healthy habits and weed out unhealthy ones. If you want to condition yourself to jog in the morning, then perhaps you should put out the running shoes the night before, where you cannot miss them when you wake up the next day. This visual cue mimics the audio cue we used to train the rats—and it could be especially effective if you reward yourself after the jog. Do this on enough mornings, and your brain might develop the chunking pattern that you want. Alternatively, if you want to forgo the candy dish, you could remove it from the living room or office—eliminating the cue.

Changing habits might never be easy. Our experiments, however, lead us to an optimistic point of view: By learning more about how our brains establish and maintain routines, we hope we can figure out how people can coax themselves out of undesirable habits and into the ones they want.

(Excerpt from *Scientific American*, June 2014)

---

[1] Tourette's syndrome is a common neuropsychiatric disorder with onset in childhood, characterized by multiple motor tics and at least one vocal (phonic) tic. These tics characteristically wax and wane, can be suppressed temporarily, and are typically preceded by an unwanted urge or sensation in the affected muscles. Some common tics are eye blinking, coughing, throat clearing, sniffing, and facial movements.

Exercises

I. Reading Comprehension

● Section One

Directions: Answer the following questions based on the information from the text.

1. What is a habit?
2. What is the clue to uncovering the brain circuitry responsible for habit formation and maintenance?
3. How does the brain stamp a behavior as one to make a habit?
4. Can we prevent a habit from forming in the first place?
5. Is it easy to break a habit? Why or why not?

● Section Two

Directions: Write an abstract based on the text in no more than 200 words.

Abstract:

Key words:

II. Vocabulary

● Section One

Directions: Choose the explanation that is closest in meaning to the underlined part in each sentence.

1. Many of them, from brushing our teeth to driving a familiar route, simply allow us to do

Unit 6
Behavioral Science

certain things on autopilot so that our brains are not <u>overtaxed</u> by concentrating on each brushstroke and countless tiny adjustments of the steering wheel.

A. require to pay too much tax

B. make excessive demands on

C. remove forcibly from power

D. cover with a wrapping

2. New techniques are finally allowing neuroscientists to decipher the neural mechanisms that underlie our rituals, including defining our so-called habit <u>circuits</u>.

A. a track used for motor racing, horse racing, or athletics

B. a series of sporting events in which the same players regularly take part

C. the path of a current in bio-electricity

D. a series of athletic exercises performed consecutively in one training session

3. That promise <u>springs</u> from one of several surprises: That even when it seems we are acting automatically, part of our brain is dutifully monitoring our behavior.

A. originates or arises from some source

B. moves or jumps suddenly or rapidly

C. brings about the escape or release of (a prisoner)

D. rises suddenly and quickly from or as from a sitting position

4. Yet it comes with a <u>trade-off</u> that can work against us.

A. the action of buying and selling goods and services

B. the action of drawing or pulling a thing over a surface, especially a road or track

C. the action of speaking badly of someone so as to damage their reputation

D. a balance achieved between two desirable but incompatible features

5. Habits become so <u>ingrained</u> that we perform them even when we do not want to, in part because of what are called "reinforcement contingencies".

A. (of a habit, belief, or attitude) firmly fixed or established in a person

B. flattered or tried to please someone

C. taken into the body by swallowing or absorbing it

D. gathered in or together

6. Somehow the brain computes these evaluations, which <u>sculpt</u> our expectations and add or subtract value from particular courses of action.

   A. make something into a particular shape

   B. engage in a short, confused fight or struggle at close quarters

   C. move hurriedly with short quick steps

   D. sink a ship deliberately

7. To figure out how the brain <u>stamps</u> a behavior as one to make a habit, the M.I.T. lab began recording the electrical activity of small collections of neurons (brain cells) in the striatum.

   A. fixes a postage stamp on to (a letter)

   B. makes something by cutting it out with a die or mould

   C. brings down (one's foot) heavily on the ground

   D. impresses a mark on a surface using an engraved or inked block or die or other instrument

8. By monitoring our actions internally and adding a positive or negative <u>weight</u> to them, the brain reinforces specific behaviors, shifting actions from deliberate to habitual.

   A. force exerted on the mass of a body by a gravitational field

   B. ability of someone or something to influence decisions or actions

   C. a unit used for expressing how much an object weighs

   D. a heavy object being lifted or carried

9. Many people know the feeling of having worked hard to break a habit only to have it come back, full-blown, after a stressful time or after one <u>relapse</u>.

   A. a deterioration in someone's state of health after a temporary improvement

   B. the renewal of youth or vitality

   C. a mushroom with a shiny cap

   D. a thing which has survived from an earlier period

10. As Mark Twain said, "Habit is habit, and not to be flung out the window by any man, but <u>coaxed</u> downstairs one step at a time."

    A. persuaded gradually or by flattery to do something

    B. manipulated carefully into a particular shape or position

    C. provided with a layer or covering of something

    D. made rough

Unit 6
Behavioral Science

• **Section Two**

Directions: There are two or three meanings for each semi-technical term underlined in the following sentences. Choose the correct one according to the context.

1. Although habits fall along different parts of the behavior spectrum (the entire range of wavelengths of electromagnetic radiation; used to classify something in terms of its position on a scale between two extreme or opposite points), they share certain core features.

2. We and others wondered what goes on in the brain's wiring (a system of wires providing electric circuits for a device or building; the structure of the nervous system or brain perceived as determining a basic or innate pattern of behavior) to cause this shift and whether we could interrupt it.

3. Somehow the brain computes these evaluations, which sculpt our expectations and add or subtract (take away from another to calculate the difference; take away something from something else so as to decrease the size, number, or amount) value from particular courses of action.

4. He and his colleagues taught lab rats in a test box to press a lever (means of exerting pressure on someone to act in a particular way; a projecting arm or handle that is moved to operate a mechanism) to receive a food treat as a reward.

5. When the animals had learned this task well and were back in their cages, the experimenters "devalued" the reward, either by letting the rats eat the reward to the point of oversatiation or by giving them a drug that produced mild nausea (a feeling of sickness with an inclination to vomit; revulsion) after the reward was eaten.

6. When the rats were first learning the maze, neurons in the motor-control (a machine that supplies motive power for a vehicle or for some other device with moving parts; relating to muscular movement or the nerves activating it) part of the striatum were active the whole time the rats were running.

7. What seemed to be happening was that the striatal cells were malleable (able to be hammered out of shape without breaking or cracking; easily influenced) and could help package movements together while leaving relatively few "expert cells" to handle the details of the behavior.

8. When Russian scientist Ivan Pavlov studied this phenomenon in dogs many years ago, he concluded that animals never forget deeply conditioned (made fit or healthy; being trained or accustomed to behave in a certain way or to accept certain circumstances) behaviors such as habits.
9. Yet remarkably, we can toggle (switch from one feature or state to another; fasten with a peg or crosspiece) the habits on and off by manipulating a tiny part of the neocortex during the actual behavior.
10. It is also possible that our expanding knowledge could even help people at the severe end of the habit spectrum, providing clues for how to treat obsessive-compulsive disorder (a state of confusion; the disruption of peaceful and law-abiding behavior; a disease or abnormal condition), Tourette's syndrome, fear or post-traumatic stress disorder.

● **Section Three**

Directions: Match the Chinese terms with their English equivalents.

1. 精神分裂症   A. brain circuitry
2. 基底神经节   B. texting
3. 纹状体   C. tweeting
4. 光遗传学   D. infralimbic cortex
5. 大脑回路   E. schizophrenia
6. 发短信   F. basal ganglia
7. 发微博   G. striatum
8. 神经元   H. lever
9. 新大脑皮层   I. chunking pattern
10. 创伤后应激障碍   J. optogenetics
11. 强迫症   K. tourette syndrome
12. 多发性抽动症   L. obsessive-compulsive disorder
13. 下边缘皮质   M. post-traumatic stress disorder
14. 记忆群组   N. neocortex
15. 操纵杆   O. neuron

III. Questions for Discussion

Directions: Work in groups and discuss the following questions.

1. Can you imagine another possible mode of brain's stamping?
2. What efforts have people made for breaking habits?
3. Are you willing to take drugs to help you control habits? Why or why not?

Part II Extensive Reading

Text B

Baby Talk

Every infant is a natural-born linguist capable of mastering any of the world's 7,000 languages like a native.

By Patricia K. Kuhl

An infant child possesses an amazing, and fleeting, gift: the ability to master a language quickly. At six months, the child can learn the sounds that make up English words and, if also exposed to Quechua and Tagalog, he or she can pick up the unique acoustic properties of those languages, too. By age three, a toddler can converse with a parent, a playmate or a stranger.

I still marvel, after four decades of studying child development, how a child can go from random babbling to speaking fully articulated words and sentences just a few years later—a mastery that occurs more quickly than any complex skill acquired during the course of a lifetime. Only in the past few years have neuroscientists begun to get a picture of what is happening in a baby's brain during this learning process that takes the child from gurgling newborn to a wonderfully engaging youngster.

At birth, the infant brain can perceive the full set of 800 or so sounds, called phonemes, that can be strung together to form all the words in every language of the world. During the second half of the first year, our research shows, a mysterious door opens in the child's brain. He or she enters a "sensitive period", as neuroscientists call it, during which the infant brain is ready to receive the first basic lessons in the magic of language.

The time when a youngster's brain is most open to learning the sounds of a native tongue begins at six months for vowels and at nine months for consonants. It appears that the sensitive period lasts for only a few months but is extended for children exposed to sounds of a second language. A child can still pick up a second language with a fair degree of fluency until age seven.

The built-in capacity for language is not by itself enough to get a baby past the first utterances of "Mama" and "Dada". Gaining mastery of the most important of all social skills is helped along by countless hours listening to parents speak the silly vernacular of "parentese". Its exaggerated inflections—"You're a preettee babbee"—serve the unfrivolous purpose of furnishing daily lessons in the intonations and cadences of the baby's native tongue. Our work puts to rest the age-old debates about whether genes or the environment prevails during early language development. They both play starring roles.

Knowledge of early language development has now reached a level of sophistication that is enabling psychologists and physicians to fashion new tools to help children with learning difficulties. Studies have begun to lay the groundwork for using recordings of brain waves to determine whether a child's language abilities are developing normally or whether an infant may be at risk for autism, attention deficit or other disorders.

• The Statistics of Baby Talk

The reason we can contemplate a test for language development is that we have begun to understand how babies absorb language with seeming ease. My laboratory and others have shown that infants use two distinct learning mechanisms at the earliest stages of language acquisition: one that recognizes sound through mental computation and another that requires intense social immersion.

To learn to speak, infants have to know which phonemes make up the words they hear all around them. They need to discriminate which 40 or so, out of all 800, phonemes they need to learn to speak words in their own language. This task requires detecting subtle differences in spoken sound. A change in a single consonant can alter the meaning of a word—"bat" to "pat", for instance. And a simple vowel like "ah" varies widely when spoken by different people at different speaking rates and in different contexts—"Bach" versus "rock". Extreme variation in phonemes is why Apple's Siri still does not work flawlessly.

My work and that of Jessica Maye, then at Northwestern University, and her colleagues have shown that statistical patterns—the frequency with which sounds occur—play a critical role in helping infants learn which phonemes are most important. Children between eight and 10 months of age still do not understand spoken words. Yet they are highly sensitive to how often phonemes occur—what statisticians call distributional frequencies. The most important phonemes in a given language are the ones spoken most.

The statistical frequency of particular sounds affects the infant brain. In one study of infants in Seattle and Stockholm, we monitored their perception of vowel sounds at six months and demonstrated that each group had already begun to focus in on the vowels spoken in their native language. The culture of the spoken word had already pervaded and affected how the baby's brain perceived sounds.

What exactly was going on here? Maye has shown that the brain at this age has the requisite plasticity to change how infants perceive sounds. It appears that learning sounds in the second half of the first year establishes connections in the brain for one's native tongue but not for other languages, unless a child is exposed to multiple languages during that period.

Later in childhood, and particularly as an adult, listening to a new language does not produce such dramatic results—a traveler to France or Japan can hear the statistical distributions of sounds from another language, but the brain is not altered by the experience. That is why it is so difficult to pick up a second language later on.

A second form of statistical learning lets infants recognize whole words. As adults, we can distinguish where one word ends and the next begins. But the ability to isolate words from the stream of speech requires complex mental processing. Spoken speech arrives at the ear as a continuous stream of sound that lacks the separations found between written words.

Jenny Saffran, now at the University of Wisconsin–Madison, and her colleagues—Richard Aslin of the University of Rochester and Elissa Newport, now at Georgetown University—were the first to discover that a baby uses statistical learning to grasp the sounds of whole words. In the mid-1990s Saffran's group published evidence that eight-month-old infants can learn wordlike units based on the probability that one syllable follows another. Take the phrase "pretty baby". The syllable "pre" is more likely to be heard with "ty" than to accompany another syllable like "ba". In the experiment, Saffran had babies listen to streams of computer-synthesized nonsense words that contained syllables, some of which occurred together more often than others. The babies' ability to focus on syllables that coincide in the made-up language let them identify likely words.

The discovery of babies' statistical-learning abilities in the 1990s generated a great deal of excitement because it offered a theory of language learning beyond the prevailing idea that a child learns only because of parental conditioning and affirmations of whether a word is right or wrong. Infant learning occurs before parents realize that it is taking place. Further tests in my lab, however, produced a significant new finding that lends an important caveat to this story: The statistical-learning process does not require passive listening alone.

● **Baby Meet and Greet**

In our work, we discovered that infants need to be more than just computational geniuses processing clever neural algorithms. In 2003 we published the results of experiments in which nine-month-old infants from Seattle were exposed to Mandarin Chinese. We wanted to know whether infants' statistical-learning abilities would allow them to learn Mandarin phonemes.

In groups of two or three, the nine-month-olds listened to Mandarin native speakers while their teachers played on the floor with them, using books and toys. Two additional groups were also exposed to Mandarin. But one watched a video of Mandarin being spoken. Another listened to an audio recording. A fourth group, run as a control, heard no Mandarin at all but instead listened to U.S. graduate students speaking English while playing with the children with the same books and toys. All of this happened during 12 sessions that took place over the course of a month.

Infants from all four groups returned to the lab for psychological tests and brain monitoring to gauge their ability to single out Mandarin phonemes. Only the group exposed to Chinese from live speakers learned to pick up the foreign phonemes. Infants who were exposed to Mandarin by television or audio did not learn at all. Their ability to discriminate phonemes matched infants in the control group, who, as expected, performed no better than before the experiment.

The study provided evidence that learning for the infant brain is not a passive process. It requires human interaction—a necessity that I call "social gating". This hypothesis can even be extended to explain the way many species learn to communicate. The experience of a young child learning to talk, in fact, resembles the way birds learn song.

I worked earlier with the late Allison Doupe of the University of California, San Francisco, to compare baby and bird learning. We found that for both children and zebra finches, social experience in the early months of life was essential. Both human and bird babies immerse themselves in listening to their elders, and they store memories of the sounds they hear. These recollections condition the brain's motor areas to produce sounds that match those heard frequently in the larger social community in which they were being raised.

Exactly how social context contributes to the learning of a language in humans is still an open question. I have suggested, though, that parents and other adults provide both motivation and necessary information to help babies learn. The motivational component is driven by the brain's reward systems—and, in particular, brain areas that use the neurotransmitter dopamine during social interaction. Work in my lab has already shown that babies learn better in the presence of other babies—we are currently engaged in studies that explain why this is the case.

Babies who gaze into their parents' eyes also receive key social cues that help to speed the next stage of language learning—the understanding of the meaning of actual words. Andrew Meltzoff of the University of Washington has shown that young children who follow the direction of an adult's gaze pick up more vocabulary in the first two years of life than children who do not track these eye movements. The connection between looking and talking makes perfect sense and provides some explanation of why simply watching an instructional video is not good enough.

In the group that received live lessons, infants could see when the Mandarin teacher glanced at an object while naming it, a subtle action that tied together the word with the object named. In a paper published in July, we also showed that as a Spanish tutor holds up new toys and talks about them, infants who look back and forth between the tutor and the toy, instead of just focusing on one or the other, learn the phonemes as well as words used during the study session. This example is an illustration of my theory that infants' social skills enable—or "gate"—language learning.

These ideas about the social component of early language learning may also explain some of the difficulties encountered by infants who go on to develop disorders such as autism. Children with autism lack basic interest in speaking. Instead they fixate on inanimate objects and fail to pay attention to social cues so essential in language learning.

- **Say, "Hiiiii!"**

An infant's ability to learn to speak depends not only on being able to listen to adults but also on the manner in which grown-ups talk to the child. Whether in Dhaka, Paris, Riga or the Tulalip Indian Reservation near Seattle, researchers who listen to people talk to a child have learned one simple truth: An adult speaks to a child differently than to other adults. Cultural ethnographers and linguists have dubbed it "baby talk", and it turns up in most cultures. At first, it was unknown whether baby talk might hinder language learning. Numerous studies, however, have shown that motherese or parentese, the revisionist name for baby talk, actually helps an infant learn.

My lab has looked at the specific sounds of parentese that intrigue infants: the higher pitch, slower tempo and exaggerated intonation. When given a choice, infants will choose to listen to short audio clips of parentese instead of recordings of the same mothers speaking to other adults. The high-pitched tone seems to act as an acoustic hook for infants that captures and holds their attention.

Parentese exaggerates differences between sounds—one phoneme can be easily discriminated from another. Our studies show that exaggerated speech most likely helps infants as they commit these sounds to memory. In a recent study by my group, Nairán Ramírez-Esparza, now at the University of Connecticut, had infants wear high-fidelity miniature tape recorders fitted into lightweight vests worn at home throughout the day. The recordings let us enter the children's auditory world and showed that if their parents spoke to them in parentese at that age, then one year later these infants had learned more than twice the number of words as those whose parents did not use the baby vernacular as frequently.

- **Signatures of Learning**

Brain scientists who study child development are becoming excited about the possibility of using our growing knowledge of early development to identify signatures of brain activity, known as biomarkers, that provide clues that a child may be running into difficulty in learning language. In a recent study in my lab, two-year-old children with autism spectrum disorder listened to both known and unfamiliar words while we monitored their brain's electrical activity when they heard these words.

We found the degree to which a particular pattern of brain waves was present in response to known words predicted the child's future language and cognitive abilities, at ages four and six. These measurements assessed the child's success at learning from other people. They show that if a youngster has the ability to learn words socially, it bodes well for learning in general.

The prospect for being able to measure an infant or toddler's cognitive development is improving because of the availability of new tools to judge their ability to detect sounds. My research group has begun to use magnetoencephalography❶ (MEG), a safe and noninvasive imaging technology, to demonstrate how the brain responds to speech. The machine contains 306 SQUID (superconducting quantum interference device) sensors placed within an apparatus that looks like a hair dryer. When the infant sits in it, the sensors measure tiny magnetic fields that indicate specific neurons firing in the baby's brain as the child listens to speech. We have already demonstrated with MEG that there is a critical time window in which babies seem to be going through mental rehearsals to prepare to speak their native language.

MEG is too expensive and difficult to use in a neighborhood medical clinic. But

---

❶ Magnetoencephalography (MEG) is a functional neuroimaging technique for mapping brain activity by recording magnetic fields produced by electrical currents occurring naturally in the brain, using very sensitive magnetometers. Applications of MEG include basic research into perceptual and cognitive brain processes, localizing regions affected by pathology before surgical removal, determining the function of various parts of the brain, and neuro feedback.

these studies pave the way by identifying biomarkers that will eventually be measured with portable and inexpensive sensors that can be used outside a university lab.

If reliable biomarkers for language learning can be identified, they should help determine whether children are developing normally or at risk for early-life, language-related disabilities, including autism spectrum disorder, dyslexia, fragile X syndrome and other disorders. By understanding the brain's uniquely human capacity for language—and when exactly it is possible to shape it—we may be able to administer therapies early enough to change the future course of a child's life.

(Excerpt from *Scientific American*, November 2015)

Exercises

I. Translate the following technical terms into English.

1. 音素 _____
2. 辅音 _____
3. 元音 _____
4. 父母语 _____
5. 自闭症 _____
6. 敏感期 _____
7. 语调 _____
8. 思维处理 _____
9. 音调变化 _____
10. 神经算法 _____
11. 注意力缺失 _____
12. 社会门控 _____
13. 多巴胺神经递质 _____
14. 分布频率 _____
15. 生物标记 _____
16. 阅读障碍 _____
17. 学习机制 _____
18. 统计模式 _____
19. 方言 _____
20. 语言习得 _____

II. Translate the following paragraphs into Chinese.

My lab has looked at the specific sounds of parentese that intrigue infants: the higher pitch, slower tempo and exaggerated intonation. When given a choice, infants will choose to listen to short audio clips of parentese instead of recordings of the same mothers speaking to other adults. The high-pitched tone seems to act as an acoustic hook for infants that captures and holds their attention.

Parentese exaggerates differences between sounds—one phoneme can be easily discriminated from another. Our studies show that exaggerated speech most likely helps infants as they commit these sounds to memory. In a recent study by my group, Nairán Ramírez-Esparza, now at the University of Connecticut, had infants wear high-fidelity miniature tape recorders fitted into lightweight vests worn at home throughout the day. The recordings let us enter the children's auditory world and showed that if their parents spoke to them in parentese at that age, then one year later these infants had learned more than twice the number of words as those whose parents did not use the baby vernacular as frequently.

Text C

Conquer Yourself, Conquer the World

Self-control is not just a puritanical[1] virtue. It is a key psychological trait that breeds success at work and play—and in overcoming life's hardships.

By Roy F. Baumeister

The ability to regulate our impulses and desires is indispensable to success in living and working with others. People with good control over their thought processes, emotions and behaviors not only flourish in school and in their jobs but are also healthier, wealthier and more popular. And they have better intimate relationships and are more trusted by others. What is more, they are less likely to go astray by getting arrested, becoming addicted to drugs or experiencing unplanned pregnancies. They even live longer. Brazilian writer Paulo Coelho summed up these benefits in one of his novels: "If you conquer yourself, then you will conquer the world."

Self-control is another name for changing ourselves—and it is by far the most critical way we have of adapting to our environment. Given that most of us lack the

---

[1] Puritanical describes someone who has very strict moral principles, and often tries to make other people behave in a more moral way.

Unit 6
Behavioral Science

kingly power to command others to do our bidding and that we need to enlist the cooperation of others to survive, the ability to restrain aggression, greed and sexual impulses becomes a necessity.

Social psychologists' appreciation of the importance of self-control reflects a shift in perspective. Thirty years ago many of them mistakenly regarded cultivation of self-esteem as a panacea for personal problems and social ills—an honest mistake❶. High self-esteem is associated with doing well in life, so it was reasonable to assume that a boost would improve people's lives.

When analyzed more closely, the data suggested that self-esteem does not itself lead to success. It is less a cause than an effect. When researchers tracked students over long periods, they found that getting good grades results in better self-esteem later. But having higher self-esteem does not produce stellar report cards. Self-control, however, is the real deal.

Experiments on self-control began in the 1960s with pioneering studies of delaying gratification conducted by Walter Mischel, now at Columbia University. Using a procedure that came to be dubbed the "marshmallow test", he offered children a choice between immediately getting the white, cylindrical candies (or another of their favorite treats) or else receiving a couple of those same sweets if they could only wait for a while. More than a decade after these early experiments were published, Mischel and his colleagues tracked down the children, by then young adults, and did so again as they entered middle age. The ones who had the most success at resisting temptation at age four went on to be the most successful as adults.

Recognizing the requirement of self-control for well-being, I and others have set about probing the psychological and biological processes underlying it. The findings indicate that the act of opting not to express anger or of choosing to forgo a marshmallow is akin to drawing on a store of energy that gets you through mile 26 of a marathon. What psychologists have learned about self-control in recent studies may even provide new ideas for treating the seemingly intractable challenges

---

❶ Honest mistake is something that you did wrong because you didn't know any better. When you call something "an honest mistake", it suggests that people shouldn't get angry at the person who made the mistake. You can use this phrase to reassure someone who's done something wrong.

of drug and alcohol addictions.

● Mental Muscle Building

I have spent a quarter of a century doing laboratory studies on self-control with an endlessly fascinating stream of creative colleagues. Over that time, I have come to the conclusion that self-control, which might also be referred to as self-regulation or willpower, works something like a muscle does. In particular, it seems to "tire" after a workout.

We coined the term "ego depletion[1]" to label the state of diminished willpower that follows from expending psychic energy on self-control, be it resisting temptation or forcing oneself to make tough decisions. The term was chosen as an homage to Sigmund Freud, who proposed that the self consists partly of a well of energy. His vague theories about how this energy worked are now mostly obsolete, but he did recognize that some form of psychic energy explains our behavior. Cast aside for decades, this idea reemerged when our experiments found that self-control operates as a mental muscle of sorts, a muscle in which energy stores get depleted with use.

Two other lines of research have extended the muscle analogy. Experiments by Mark Muraven of the University at Albany and his colleagues have shown that after exertion, willpower has not entirely vanished. Rather the body seems to be conserving energy; if an important challenge or opportunity arises, more self-control can be tapped. This finding parallels what happens with physical muscles. As muscles begin to tire, athletes cut back on exertion to conserve remaining energy and strength. But they can marshal concerted effort if needed, calling on reserves for a sprint to the finish.

Muscles do not just become fatigued; they increase in strength when used regularly. Self-control can also strengthen with practice. In several studies, volunteers were assigned for a two-week period to change how they speak—avoiding curse words, using complete sentences, and saying "yes" and "no" instead of "yeah" or "nope". In another program, subjects were simply asked to improve their posture—

---

[1] Ego depletion refers to the idea that self-control or willpower draws upon a limited pool of mental resources that can be used up.

sitting or standing up straight. After the exercises were completed, we evaluated the subjects' self-control using lab tests. Those who had practiced the earlier exercises performed significantly better than a control group that had not had to clean up their language or sit up straight. It has occurred to us from these studies that the Victorian notion of "building character" seems to have some scientific validity. Exerting self-control on a regular basis appears to build up a person's capacity to call on more of this character trait in a pinch.

Matt Gailliot, then a graduate student, wondered whether we could extend the observation that willpower becomes depleted when someone resists temptation. What about the opposite case? Would indulging in temptation actually strengthen willpower?

I had my doubts, but I encouraged Gailliot to pursue the question, which we informally called the "Mardi Gras❶ theory", in reference to the Christian tradition of indulging in sinful impulses in preparation for a period of self-denial during Lent. First we sapped people's self-control by requiring them to mentally suppress the forbidden thought of a white bear. Then we randomly assigned some of the participants to drink a delicious ice cream milk shake before they took a disguised test of willpower that consisted of searching a matrix of numbers for a particular sequence.

The folks who drank the shake persevered longer on the test than those who got nothing. This apparent victory for the Mardi Gras theory was soon undercut by another result that involved an additional control group. One of the groups, as before, received nothing to drink before the test and, as expected, did badly on that test. The other group drank a milk shake that did not taste good; it contained unsweetened half-and-half rather than ice cream, so it was basically a large, unappetizing glass of dairy glop. Unfortunately for Gailliot's theory, the half-and-half group also did better than the unfed subjects. Gailliot was initially glum because the experiment seemed a bust. But as we talked, another thought occurred to us: If

---

❶ Mardi Gras is also called Shrove Tuesday, or Fat Tuesday. In English, it refers to events of the Carnival celebrations, beginning on or after the Christian feasts of the Epiphany (Three Kings Day) and culminating on the day before Ash Wednesday. Mardi Gras reflects the practice of the last night of eating richer, fatty foods before the ritual fasting of the Lenten season.

it was not the pleasure of indulgence that restored willpower, could it have been the calories?

We started reading up on glucose, the sugar in the blood-stream that provides energy to bodily tissues, including the brain, the seat of self-control. We ran a large series of studies and came up with two supportive findings that have stood the test of time. One showed that when blood glucose is low, self-control suffers, often substantially. This pattern, by the way, gives credence to the oft-heard complaint that a person is having difficulty functioning because of "low blood sugar"—a conclusion that also jibes with studies from nutritionists.

The other meaningful finding confirmed that a dose of glucose administered just before self-control is beginning to flag helps to restore the needed willpower to press ahead. These results strongly suggest that willpower is, indeed, more than a metaphor. Further, if exerting self-control diminishes willpower and the energy needed to sustain it, then the remaining energy can be conserved by cutting back on further demands for self-control.

A third result did not hold up. We found in one study that blood glucose levels drop during a task that requires self-control. Such a finding would be consistent with the idea that exerting willpower uses up glucose. But we could not replicate the pattern reliably in later tests. Some studies from other labs have shown, however, that the brain uses more glucose when exerting greater effort—which makes sense, after all, given that it is the brain that controls self-restraint.

● **A Challenge to Our Ideas**

Like many scientific theories, our muscle model of self-control has evolved as other researchers have gotten into the act. Some have tried to build on what we have done, and others have wanted to dismantle or challenge our work. These new findings—and the debates they have engendered—have helped flesh out our understanding of self-control.

One contentious issue has been whether the brain really runs out of fuel for willpower. Like us, other investigators have confirmed that self-control is impaired when blood glucose is low, a physiological state that affects both body and brain.

Some researchers have argued that the human body has extensive reserves of glucose that could be drawn on if an amount allotted to willpower got used up. Compounding the skepticism over our notion of energy depletion, the brain's glucose consumption does not fluctuate much—still, it does change some.

All these points are well taken. It is possible, though, that exercise of self-control does not necessarily lead directly to the exhaustion of glucose and that when the body senses that available glucose is running low, it makes adjustments to direct the sugar to where it is needed most. In that case, we would still be correct in thinking that willpower is a precious resource—one that needs to be conserved.

Another critique suggests that any willpower deficit can be overcome by just putting people with declining reserves into circumstances that cause them to call up additional resolve. Studies have shown that assigning people to a position of power and leadership—or even paying them to try harder—makes them continue to show good self-control even in situations where their energy should be depleted by prior exertion of willpower.

This research raises the possibility that willpower is all in your head. No resource is actually depleted, but people simply lose motivation to work hard. It can also mean that when willpower declines, you can still exert effective self-control if doing so is critical.

Our energy-allocation theory does not entirely disagree with the view that people can draw on spare resources for a time. If your willpower is slightly depleted, your body may naturally seek to conserve what remains—but you can still suck it up and perform well if the situation warrants. Tired athletes conserve their energy for the winnable points and the crucial, decisive moments. Ego-depleted people do the same with willpower.

In our own studies, we have found that people who believe in unlimited willpower tap into existing reserves to increase blood glucose levels when the sugar should have otherwise been depleted. The story, though, grows a little more complicated when examined more closely.

A crucial test came when people were not just slightly depleted but continued exercising self-control until serious fatigue could no longer be ignored. Kathleen Vohs of the University of Minnesota, Sarah Ainsworth and others had shown that cash incentives or leadership responsibilities enable people to sustain self-control even when their willpower is depleted. But these various studies then initiated a grueling series of exercises, which showed that depletion worsened, and self-control started to diminish. Crucially, those who had been led to believe in unlimited willpower actually did worse than others. That belief had been helpful at first, but in the long run it backfired.

Self-control, it seems, can be maintained—but not indefinitely. You just become more willing to spend from your reserves. Eventually a limit is reached. The illusion of endless self-control is tantamount to believing that a bank account has infinite funds. At the beginning, you may spend freely, but ultimately you seriously risk running out of money.

- **Can You Will Away an Addiction?**

Recent studies have revealed newly discovered areas in which self-control plays a pivotal role. Some of these findings overturn prevailing ideas about various forms of addiction. A widely held view suggests that cravings for drugs, alcohol or cigarettes take over an addict's life and that quitting is impossible without complex medical treatments or at least a firm commitment to a 12-step-like program. Alan I. Leshner, former director of the National Institute on Drug Abuse and now CEO of the American Association for the Advancement of Science, has asserted that addiction is a "brain disease". As he put it, a user may take a puff or inject a substance voluntarily, but at some point, a switch in the brain is thrown. Substance abuse becomes involuntary, and the compulsion lingers even when the addict earnestly desires to quit. Willpower and volition disappear once addiction takes hold.

New findings indicate, however, that any brain changes occurring in addicts do not lead to a loss of control over one's actions; often these people have the power to choose whether to give in to a craving or resist.

More specifically, addiction does not bring about changes in a brain area essential for self-control that governs movement—that is, the motor cortex, where actions are initiated. As addiction grows, the decision to grab the pipe does not suddenly become involuntary. Instead addiction brings on a slow and insidious change in desire. Heroin or cigarettes evoke pleasant feelings that develop into a longing for these substances. The addict can resist for a time but gives in at some point, perhaps sooner rather than later, and must thwart the desire again and again. This and other findings indicate that the addict experiences an intermittent stream of one mild urge after another.

The controversy about whether addicts are still in control will likely persist. Arguments from politicians, drug counselors and others help to sustain the myth that addiction is rooted in overwhelming, uncontrollable urges. Many addicts themselves favor this viewpoint because it exonerates them from personal responsibility.

Psychologists differ as to whether self-control can be an effective anti-addiction medicine. A survey in the U.K. found that addiction-treatment counselors who worked as volunteers tended to think that addicts can regulate their impulses. But those who received compensation for their work preferred to think that addicts are helpless and cannot get better without expert help. This argument is not intended to suggest that clinicians are in it only for the money. But when a controversy arises, financial incentives probably make it easier for people to endorse evidence that goes along with their own interests and to spot flaws in counterarguments.

Another addiction myth holds that cravings grow more acute only when quitting an addictive substance. A clever study by Michael Sayette of the University of Pittsburgh and his colleagues demonstrated that smokers believed that their desire would increase steadily over time, especially if they were told they could not light up.

The study also showed that these beliefs were wrong. Other studies have found that when a smoker quits, the desire to smoke goes down immediately and mostly stays in abeyance.

- **Addiction Is for the Strong-Willed**

The idea that quitting an addiction requires willpower makes sense to most people. But until recently, few have considered that starting a drug habit and staying addicted also require self-control. Most of us do not really like the first taste of beer or the first puff of a cigarette. To sustain an addiction over a long period, a user must expend a substantial amount of energy to ensure that a habit does not interfere with work, family and relationships.

Consider smoking. So many restrictions exist today that smokers need to craft elaborate plans to sneak a cigarette. When my former university introduced rules prohibiting professors from smoking in their offices, one colleague struggled heroically to comply. I will not soon forget the sight of her heading out of the building into a Cleveland snowstorm, while holding her tiny baby in her arms, on the way to light up.

Just think about how much self-control she had to muster. First, she had to plan when she would find breaks between classes, appointments and meetings—and where she would go to not violate campus smoking restrictions. Then she had to dress herself and the baby warmly. She also had to remember her cigarette pack and lighter on the way out into the storm.

Scientists who favor the view that addicts have little self-control might have expected a different initial outcome—high self-control types would alter their behavior in response to the ban, whereas poor self-regulators would keep right on smoking. And they might explain the fact that we found the opposite result by reasoning that people with low self-control needed the strong push from the law to get them over the hump—and interpret the subsequent relapse by suggesting that over time the threat of a legal cudgel somehow faded. But the explanation for the results appears to be related to the addicts' need to draw on reserves of willpower to preserve their habit.

A number of studies have shown that addicts seem able to consistently plan and execute intricate strategies to maintain heroin or cigarette habits—habits that researchers, clinicians and even users themselves once thought to be unshakable. These findings provide a new perspective on addiction. The possibility exists that these groups may be able to redirect the same sustained willfulness they use to procure a drink or fix toward kicking their habits. But this idea also raises a new set of issues.

A therapist may have difficulty convincing an addict that he or she has taken the wrong path if that person sees nothing permanently damaging with having a few drinks or popping pain-killers while continuing to fulfill responsibilities at home and work. This new insight into the nature of addiction provides further evidence of the extent that self-control can influence our behaviors in myriad ways—and how it may even, perhaps counterintuitively, enable us to persist in adhering to self-destructive habits. It demonstrates, once again, that our ability to control our emotions and desires lets us manage, for good or bad, the endless challenge of adapting to the world around us.

(Excerpt from *Scientific American*, April 2015)

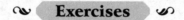

Reading Comprehension

Directions: Answer the following questions based on the information from the text.

1. What is the relationship between self-esteem and self-control?
2. Why does self-control work as a muscle according to the author?
3. In what way is addiction a "brain disease"?
4. What does "ego depletion" mean in the text?
5. How is the mental muscle built?

Unit 7
Material Science

导读

本单元的文章涉及材料科学领域中的几个新宠。Text A介绍了光伏领域中钙钛矿太阳能电池的工作原理、性能及其优点和缺点，并详细描述了科学家们在钙钛矿太阳能电池研发道路上的种种尝试和努力。Text B在分析石墨烯这种新型材料的优缺点的基础上，展望了未来能克服石墨烯缺点的新型材料的开发与应用前景。Text C阐述了电池的工作原理，以及传统电池存在的问题，最后详细介绍了研究人员设计可控电极结构以提高电池能源利用效率的过程。

Part I Intensive Reading

Text A

Outshining Silicon

An upstart material—perovskite—could finally make solar cells that are cheaper and more efficient than the prevailing silicon technology.

By Varun Sivarm, Samuel D.Strank and Henry J. Snaith

Graduate student Michael Lee jotted down a list of chemical ingredients as he was sitting in a bar in Japan. Earlier that day scientists at Toin University of Yokohama had generously shared their groundbreaking recipe for making solar cells from a new material called perovskite rather than the usual silicon. The cells were only 3.8 percent efficient in converting sunlight to electricity, so the world had not taken notice. But Lee was inspired and made a series of tweaks to the recipe. The changes yielded the first perovskite cell to surpass 10 percent efficiency. His invention sparked the clean-energy equivalent of an oil rush, as researchers worldwide raced to push perovskite cells even higher.

The latest record, set at 20.1 percent in November 2014, marked a five-fold increase in efficiency in just three years. For comparison, after decades of development state-of-the-art silicon solar cells have plateaued at about 25 percent, a target that perovskite researchers like us have squarely in our sights. We are also anticipating a commercial debut, perhaps through a spin-off company such as Oxford Photovoltaics[1], which one of us (Snaith) co-founded.

Perovskites are tantalizing for several reasons. The ingredients are abundant, and researchers can combine them easily and inexpensively, at low temperature, into thin films that have a highly crystalline structure similar to that achieved in silicon wafers after costly, high-temperature processing. Rolls of perovskite film that are thin and flexible, instead of thick and rigid like silicon wafers, could one day be rapidly spooled from a special printer to make lightweight, bendable, and even colorful solar sheets and coatings.

Still, to challenge silicon's dominance, perovskite cells will have to overcome some significant hurdles. The prototypes today are only as large as a fingernail; researchers have to find ways to make them much bigger if the technology is to compete with silicon panels. They also have to greatly improve the safety and long-term stability of the cells—an uphill battle.

• Winning the Efficiency Race

Today the best silicon cells are 25.6 percent efficient. Why can't solar cells convert 100 percent of the sun's light energy? And why should perovskites be able to surpass the silicon record?

The answers to these questions are found in the excitable and errant electron. When a solar cell is in the dark, electrons in the material stay bound to their respective atoms. No electricity flows. But when sunlight strikes a cell, it can liberate some of the electrons. Infused with energy, the "excited" electrons careen drunkenly through the crystal lattice of the cell until they either exit one end of the cell—

---

[1] Oxford Photovoltaics was founded in 2010 as a spin-out from the University of Oxford to commercialize a new technology for thin-film solar cells and the first mover, leading the commercialization of perovskite technology.

whisked away by an electrode as useful current—or run into an obstacle or a trap, losing their energy in the form of waste heat.

The higher the crystal quality, the fewer defects there are to derail the electron's journey. Silicon cells are typically heated to as much as 900 degrees Celsius to remove defects. Perovskites are largely free of such defects even though they are processed at much lower temperatures, around 100 degrees C. As a result, electrons excited by light are just as successful in exiting perovskite cells, and they are unlikely to lose as much energy along the way when colliding with obstacles. Because the electrical power of a cell is the product of the flow of electrons exiting that cell (the current) and the energy that those electrons carry (the voltage), the efficiency of perovskites can rival silicon, with much less processing effort.

But there is a ceiling to how much of the sunlight's energy a solar cell made of semiconductors such as silicon and perovskites can convert into electrical power. That is primarily because of a property of semiconductors called the bandgap—a minimum level of energy needed to liberate electrons. Sunlight includes all wavelengths of light, but only certain wavelengths exceed the energy bandgap. Other wavelengths will simply pass through the material, doing nothing.

The bandgap is different for different semiconductors, and it sets up a fundamental trade-off: the lower the bandgap, the more of the sun's spectrum a cell can absorb to excite electrons, but the lower the energy each electron will have. Because electrical power depends on both the number and energy of electrons, even a cell with the ideal bandgap can convert only around 33 percent of the sun's energy.

Silicon has a fixed bandgap that is not ideal, but it commands the solar industry because effective ways to manufacture the technology are well understood. When making perovskites, however, researchers can adjust the bandgap at will by tweaking the mix of ingredients, which raises the prospect of exceeding silicon efficiencies. Researchers can also layer different perovskites with different bandgaps on top of one another. Double-decker perovskites should be able to break through the nominal 33 percent ceiling; some projections indicate they could put 46 percent of the sun's energy to work.

Unit 7
Material Science

● **Teaching an Old Material New Tricks**

Mineralogists have known about the natural forms of perovskite in the earth's crust since the 19th century. The crystals graced a 1988 cover of this magazine when scientists thought they could form high-temperature superconductors (some work continues today). During the past two decades engineers also made experimental electronics with man-made perovskites, but they overlooked the material's potential use in solar cells.

Finally, in 2009, a group at Toin University turned a man-made version into a solar cell. The researchers dissolved selected chemicals in solution, then spun and dried that solution on a glass slide. The drying left behind a film of nanometer-scale perovskite crystals on top of the slide. This film generated electrons when it absorbed sunlight but not very well. The researchers added thin layers of material on either side of the perovskite nanocrystals to help them transfer the electrons to an external electrical circuit, supplying useful power.

The first tiny cells were only 3.8 percent efficient, and they were highly unstable, deteriorating within hours. Lee altered the perovskite's composition and replaced a problematic layer in the cell, pushing the efficiency beyond 10 percent. Another set of investigators, led jointly by Michael Grätzel and Nam-Gyu Park, made a similar advance.

The recent march to 20 percent has been driven by some clever innovations. Creating a defect-free crystalline film requires tricky deposition methods, so a group headed by Sang Il Seok devised a multistep process that forced a more orderly crystal film to drop out of the spinning solution. By optimizing processing, Seok marched through three consecutive record efficiencies in 2014, from 16.2 to 20.1 percent.

Other scientists simplified the layering of added materials; the newest perovskite cells look more like a silicon cell. In silicon's case, this design has made low-cost mass production possible. Recently perovskite researchers have also heated up the solution and the glass slide on which it is deposited, resulting in crystals that are several orders of magnitude bigger than those in the initial cells.

Scientists are devising some novel traits, too. Varying the chemical ratio can create cells that have a gentle shade of yellow or a blush of crimson. Depositing perovskite on glass in islands instead of one thin layer can create films that are opaque or transparent or degrees in between. Together these options could help architects design skylights, windows and building facades that incorporate colorful perovskite solar films. Imagine a skyscraper with perovskite-tinted windows that shade the interior from hot sunlight by converting it into electricity, reducing the cooling bill while also providing power.

● Long Road to Commercialization

Perovskites have a long way to go before they fulfill such visions. Although Korean and Australian researchers recently demonstrated printable cells that are 10 by 10 centimeters, the most efficient cells are still small prototypes. As labs and start-up companies scale up the devices, they must accomplish three prerequisites for commercialization: ensure that the cells are stable enough to produce electricity for decades, design a product that customers feel is safe to put in their homes and buildings, and satisfy critics who caution that the claims for perovskite efficiency levels are inflated.

The stability of the perovskite solar cell is arguably its Achilles' heel[1]. Perovskites can degrade rapidly because they are sensitive to moisture, so they must be encased in a watertight seal. Cells fabricated by us in an inert atmosphere and encapsulated in epoxy have performed stably for more than 1,000 hours when exposed continuously to light. Researchers at the Huazhong University of Science and Technology in China, in collaboration with Grätzel, have also reached 1,000 hours even without encapsulation, and in recently published work they have deployed test panels outdoors in Saudi Arabia to show that their design will function in real-world conditions.

The industry convention for solar panels is a 25-year warranty, however. That equates to about 54,000 hours under constant, bright sunlight. Finding an effective

---

[1] Achilles' heel is a weakness in spite of overall strength, which can actually or potentially lead to downfall. Achilles is a Greek hero, in the *Iliad* the foremost of the Greek warriors at the siege of Troy. While he was a baby his mother plunged him into the river Styx making his body invulnerable except for the heel by which she held him. After slaying Hector, he was killed by Paris who wounded him in the heel.

moisture barrier that works for that long, over a wide temperature range, is crucial. Silicon manufacturers solved the problem by laminating the cells between glass sheets. This is perfect for large, ground-based installations. But because perovskite cells can be made as films that are much lighter and more flexible than cells on glass, alternative encapsulation strategies may open up broader applications, such as veneers for walls or windows that can generate electricity. Fortunately, some progress has been made by companies trying to commercialize other flexible materials, such as the semiconductor made of copper indium gallium selenide. Its encapsulation technologies work well. Perovskites may be able to exploit the encapsulation advances.

Just as important as sealing out moisture is sealing in the cells' contents because of the tiny amount of lead added to the perovskite recipe. Lead is toxic, so the market will demand a high burden of proof that perovskite power is safe. For inspiration, researchers can again look to an alternative solar material, the only one besides silicon that has achieved significant commercial success: cadmium telluride. Manufactured by First Solar, cadmium telluride panels have been deployed around the world despite the presence of an element far more toxic than lead: cadmium. First Solar has convinced communities that its panels are so well sealed that no cadmium could escape, even in a desert wildfire at 1,000 degrees C. The panels use a glass substrate, however, which precludes the flexibility and lower weight that perovskites promise. Yet perovskite companies can learn from First Solar's success in sealing and rigorously testing products.

An encouraging development related to lead recently emerged from the M.I.T. as well: Angela Belcher and her colleagues demonstrated that lead-acid car batteries can be recycled safely, with the lead content recovered to make perovskite cells. This result could be an environmental plus.

A different route would be to eliminate the lead altogether. Both our group and another one at Northwestern University have published preliminary reports on cells that use tin instead of lead. The efficiency and stability are worse, however, because tin tends to cause the perovskite to lose its crystalline structure over time, hampering an electron's ability to get out of the cells. A major advance would be needed for tin to match lead's long-term performance.

In addition to the issues listed here, researchers have to solve a smaller, quirkier problem. Critics have claimed that the efficiency numbers for perovskite cells might be inflated because of hysteresis—a jitter in the measurement that is likely caused by charged molecules migrating from one side of a cell to the other, which could create the appearance of greater current. This ion migration is very brief, however. Scientists are looking for ways to halt it, but in the near term, there is a simple remedy: wait out the migration and measure efficiency over a longer period. In most cases, this process renders efficiency readings that are similar to quick, initial measurements, but researchers may be tempted to report the higher of the readings. We are working with investigators worldwide to standardize the measurement process so that our results meet a high standard of scrutiny.

Finally, to succeed commercially, perovskite innovators need to provide a compelling economic narrative to attract the investment dollars required for scaling up production. Although materials for perovskites are abundant and cells can be processed at low temperatures into films that roll off inexpensive equipment, perovskite solar companies should not fall into the trap of competing on silicon's terms. There is little room to undercut silicon panels because most of the cost of an installation is not related to the panels but to what is called the "balance of system❶". An average U.S. residential solar installation in 2014 was priced at $3.48 per watt of electricity-generating capacity, yet the cost of the actual solar panel was only 72 cents per watt. Even if perovskite panels achieve the dirt-cheap 10 to 20 cents per watt that researchers think is possible, the improvement would reduce the final installed price by only a small percentage.

Perovskite companies can build on those small savings, though, by devising products that beat silicon's efficiencies. A highly efficient perovskite solar panel reduces the total installed cost per watt by requiring less land or roof space and therefore less labor and equipment. An even more imaginative example of changing

---

❶ The balance of system (BOS) encompasses all components of a photovoltaic system other than the photovoltaic panels. This includes wiring, switches, a mounting system, one or many solar inverters, a battery bank and battery charger.

the rules would be to sell perovskite products for applications that silicon cannot compete in, such as films that could be integrated right into building materials for walls, roofs and windows.

● **The Hybrid Solution**

For now perovskites might have the best chance to reach the market as an ally rather than a competitor of silicon. Perovskites could literally piggyback off silicon's success, gaining entry to a $50-billion market.

An alliance could happen by adding a perovskite layer right on top of a silicon layer, creating a "tandem" solar cell. Perovskites are good at harnessing the higher-energy colors of sunlight such as blue and ultraviolet, which silicon fails to capture, generating a much higher voltage in electrons. Researchers at Stanford University and M.I.T. recently stacked a perovskite cell on top of a sealed silicon cell, raising efficiency from the silicon's original 11 to 17 percent. They also assembled a tandem cell by layering perovskite on top of unsealed silicon, creating a single structure. The combination achieved just 14 percent efficiency, but that figure could surely go up with manufacturing refinements. Based on the two experiments, the researchers sketched out a scenario by which a tandem cell made with a state-of-the-art silicon component and a state-of-the-art perovskite device, combined using clever engineering, could surpass 30 percent efficiency without any radical change in either technology.

If a tandem solar panel could reach 30 percent efficiency, the impact on the balance-of-system cost could be enormous: Only two thirds of the number of panels would be needed to produce the same amount of power as panels that are 20 percent efficient, greatly reducing the amount of roof space or land, installation materials, labor and equipment. Down the line, cheap solar coatings integrated into roofing or glazing materials could transform the entire cost structure of a solar-powered building.

● **Running in Reverse**

The quick rise of perovskite solar cells has inspired scientists and engineers to fabricate other types of prototype products that also might one day make it to market. Working with our colleagues at the University of Cambridge, we recently created light-emitting diodes (LEDs) and lasers using metal halide perovskites, which efficiently emit light (instead of absorbing it) through a process called luminescence. This turnabout is not really surprising; when run in reverse, the world's most efficient solar cell, gallium arsenide, acts as an LED. Cheap, printable LEDs and lasers could lead to intriguing applications, from large-scale lighting to medical imaging.

Perovskites make scientists feel like children in a candy shop; we have found a material whose properties fill almost every checkoff box on our wish list❶, including high efficiency, low cost, light weight, flexibility and aesthetic appeal. It will take a concerted, global effort by academia, industry and government to fully realize the potential perovskites have to move beyond the silicon era.

(Excerpt from Scientific American, July 2015)

Exercises

I. Reading Comprehension

• **Section One**

Directions: Answer the following questions based on the information from the text.

1. What does "outshining silicon" refer to in the text? What advantages does it have over silicon?
2. How does solar cell convert sunlight into electrical power? Give reasons why perovskite cells are more efficient than silicon cells.
3. How did scientists invent the first tiny perovskite cells and how did they improve the perovskite cells' performance?
4. What are the three prerequisites for the commercialization of perovskite cells? And how did researchers solve these problems?

---

❶ Fill almost every checkoff box on our wish list means meeting a list of functional requirements for a product.

5. What is the prospect of the research into perovskites-related new products? Elaborate your answer.

● **Section Two**

Directions: Write an abstract based on the text in no more than 200 words.

Abstract:
Key words:

II. Vocabulary

● **Section One**

Directions: Choose the explanation that is closest in meaning to the underlined part in each sentence.

1. After the 2011 fact-finding mission, he returned to Clarendon Laboratory, where all three of us worked, and made a series of tweaks to the recipe.

    A. sharp twists or pulls   B. fine adjustments
    C. anxieties   D. big changes

2. Researchers can also layer different perovskites with different bandgaps on top of one another. Double-decker perovskites should be able to break through the nominal 33 percent ceiling.

    A. of little importance or value
    B. named or bearing the name of a specific person
    C. very small or far below the real value or cost

D. existing, etc. in name or word only, not in fact

3. The crystals graced a 1988 cover of this magazine when scientists thought they could form high-temperature superconductors.

   A. behaved gracefully
   B. did honor to someone by one's presence
   C. was an attractive presence in or on
   D. hoped to become

4. Yet perovskite companies can learn from First Solar's success in sealing and rigorously testing products.

   A. in an extremely thorough or accurate manner
   B. physically demanding
   C. adhering strictly to a way of doing something
   D. demanding strict attention to rules

5. But there is a ceiling to how much of the sunlight's energy a solar cell made of semiconductors such as silicon and perovskites can convert into electrical power.

   A. the overhead upper surface of a room
   B. maximum altitude at which a plane can fly
   C. altitude of the lowest layer of clouds
   D. an upper limit on what is allowed

6. For comparison, after decades of development state-of-the-art silicon solar cells have plateaued at about 25 percent.

   A. reached a state of little or no change after a time of activity or progress
   B. overstepped the boundary after a time of activity or progress
   C. broken the records after a time of activity or progress
   D. increased rapidly after a time of activity or progress

7. Perovskites could literally piggyback off silicon's success, gaining entry to a $50-billion market.

   A. release something from one's back
   B. use existing work or an existing product as support
   C. avoid completely
   D. imitate

8. Finally, to succeed commercially, perovskite innovators need to provide a compelling economic narrative to attract the investment dollars required for scaling up production.

Unit 7
Material Science

   A. a spoken or written account on how to make money

   B. a survey on how to make money

   C. a project design on how to make money

   D. a professional lecture on how to make money

9. They also have to greatly improve the safety and long-term stability of the cells—an uphill battle.

   A. a battle in the hill   B. a battle on the slope of the hill

   C. a battle which lasts for a long time   D. difficult battle

10. Critics have claimed that the efficiency numbers for perovskite cells might be inflated because of hysteresis, which could create the appearance of greater current.

    A. exaggerated   B. mistaken   C. misled   C misunderstood

● **Section Two**

Directions: There are two or three meanings for each semi-technical term underlined in the following sentences. Choose the correct one according to the context.

1. Researchers can combine them easily and inexpensively, into thin films (thin coating or covering; roll or sheet of thin flexible material for use in photographs) that have a highly crystalline structure similar to that achieved in silicon wafers after costly, high-temperature processing.

2. Rolls of perovskite film that are thin and flexible, instead of thick and rigid (unable to bend or be forced out of shape; not able to be changed or adapted) like silicon wafers, could one day be rapidly spooled from a special printer to make lightweight, bendable, and even colorful solar sheets and coatings.

3. The lower the bandgap, the more of the sun's spectrum a cell can absorb to excite (cause strong feelings of enthusiasm and eagerness in; produce a magnetic field in; produce a state of increased energy or activity in) electrons, but the lower the energy each electron will have.

4. Because the electrical power of a cell is the product (something produced by nature or by man; substance obtained by chemical reaction; quantity obtained by multiplication) of the flow of electrons exiting that cell and the energy that those electrons carry, the efficiency of perovskites can rival silicon.

5. Perovskites can degrade rapidly because they are sensitive to moisture, so they must be encased in a watertight seal (a stamp affixed to a document; a finishing coat applied to exclude moisture; pelt or fur).

6. Creating a defect-free crystalline film requires tricky deposition (the act of putting something somewhere; the natural process of laying down a substance on rocks or soil; the act of removing a powerful person from a position or office) methods.

7. In most cases, this process renders efficiency readings (written material intended to be read; a datum about some physical state that is presented to a user by a meter or similar instrument) that are similar to quick, initial measurements, but researchers may be tempted to report the higher of the readings.

8. Manufactured by First Solar, cadmium telluride panels have been deployed around the world and have exceeded safety standards despite the presence of an element far more toxic than lead (first place or position; conductor conveying current from the place where it is used; a soft heavy toxic malleable metallic element).

9. Finding an effective moisture barrier (any condition that makes it difficult to make progress or to achieve an objective; anything serving to maintain separation by obstructing vision or access) that works for that long, is crucial.

10. First Solar has convinced communities that its panels are so well sealed that no cadmium could escape (break free from confinement or control; leak from a container), even in a desert wildfire at 1,000 degrees C.

- **Section Three**

Directions: Match the Chinese terms with their English equivalents.

1. 离子迁移　　　　　　　　　A. encapsulation technology
2. 透光材料　　　　　　　　　B. luminescence
3. 滞后效应　　　　　　　　　C. tandem cell
4. 带隙　　　　　　　　　　　D. light-emitting diode
5. 释放电子　　　　　　　　　E. silicon wafer
6. 晶格　　　　　　　　　　　F. perovskite solar cell
7. 玻璃基板/片　　　　　　　　G. inert atmosphere
8. 叠层电池　　　　　　　　　H. crystal lattice

9. 钙钛矿太阳能电池   I. watertight seal
10. 惰性气体环境   J. ion migration
11. 发光二极管   K. glazing material
12. 防水封装   L. to liberate electron
13. 冷光   M. bandgap
14. 封装技术   N. glass substrate
15. 硅晶片   O. hysteresis

III. Questions for Discussion

Directions: Work in groups and discuss the following questions.

1. What is the latest development about the research of perovskite cells in the world?
2. What benefits will we get from perovskite cells if researchers can solve the key technical problems of perovskite cells?
3. What do you think is the prospect of applying perovskite cells in China?

Part II Extensive Reading

Text B

Beyond Graphene

The ultrathin form of carbon has inspired other atoms-thick materials that promise even bigger technological payoffs.

By Robert F. Service

When physicists Andre Gei and Konstantin Novoselov of the University of Manchester in the United Kingdom and colleagues reported in *Science* in 2004 that they had used clear tape to peel off single atomically thin sheets of carbon atoms from a chunk of graphite, it set off a revolution in materials science that is still unfolding.

Last year, researchers around the globe published more than 15,000 papers on single-layer graphite, called graphene. Graphene is the thinnest material ever made. It's 100 times stronger than steel, a better electrical and heat conductor than copper, flexible, and largely transparent. Investigators envision a future for it in everything from the next generation of computer chips and flexible displays to batteries and fuel cells.

Yet graphene may have its biggest impact not as a wonder material in its own right, but through its offspring. For all its dazzling promise, graphene has drawbacks, especially its inability to act as a semiconductor, the keystone of microelectronics. Now, chemists and materials scientists are striving to move beyond graphene. They're synthesizing other two-dimensional sheet-like materials that promise to combine flexibility and transparency with electronic properties graphene can't match. And they are already turning some of them into thin, flexible, speedy electronic and optical devices that they hope will form the backbone of industries of the future.

In one sense, 2D materials aren't new at all. Researchers have been growing atomically thin sheets of materials since the 1960s using tools called molecular beam epitaxy (MBE) machines. But MBE machines are typically used to deposit thin layers of materials like silicon and gallium arsenide: crystalline materials whose component atoms normally prefer to bond in three dimensions. In that respect, the layers made by MBE are like a slice of cheese, a 2D version of a 3D substance.

Graphene is different. It's more like the pages in a book. Its carbon atoms form strong, covalent links❶ with other carbons in a single 2D plane, creating a hexagonal lattice that looks like miniature chicken wire. But separate planes of atoms are only loosely paired with weak bonds known as van der Waals interactions❷. As a result, the layers can slip past one another, which is why graphite is used to make the gray flaky "lead" in pencils.

The big surprise was that when researchers began to study graphene closely, they discovered it had electronic and optical properties not found in bulk graphite. "The biggest lesson is that less is different," says Yuanbo Zhang, a condensed matter physicist at Fudan University in Shanghai, China. And with that lesson, Tomanek says, "graphene brought 2D materials into the limelight."

Yet when it comes to making high-tech devices, graphene's promise dims a bit. While the most prized materials of the electronics age are semiconductors, whose conductivity can be switched on and off to generate the digital currency of 1s and 0s, graphene is more like a conducting metal.

Researchers have spent years trying to convert graphene into a semiconductor by bonding oxygen to the graphene sheets or by cutting the sheets into ribbons just a few nanometers wide. Both changes do alter graphene's electronic structure, turning it into a semiconductor. But these "solutions" brought other problems. Graphene oxide's electronic properties are strongly affected by molecules that interact with it, a foible that undermines its reliability. And the nanoribbons' electronic properties depend so critically on a ribbon's precise structure that they are hard to control.

Yet graphene opened researchers' eyes to a new world of flatland electronics.

---

❶ Covalent link, also called covalent bond, refers to a chemical bond that involves sharing a pair of electrons between atoms in a molecule, especially the sharing of a pair of electrons by two adjacent atoms.

❷ van der Waals interactions, named after Dutch scientist Johannes Diderik van der Waals, are the residual attractive or repulsive forces between molecules or atomic groups that do not arise from covalent bonds, nor ionic bonds.

They saw that similar materials might have novel optical and electrical properties. And because 2D sheets are so thin and mostly transparent, they offered the prospect of creating flexible and transparent electronics that could produce see-through displays of the sort dreamed up years ago by Hollywood. Since then, researchers have been surveying that landscape for richer treasures.

By looking for materials that naturally form 2D sheets and finding ways of stabilizing sheets of atoms that normally want to form a 3D architecture, materials scientists have already come up with dozens of new 2D materials, and many more are likely to follow. They've engineered single-layer silicon (known as silicene), single-layer germanium (germanene), and single-layer tin (stanene). They've created an insulator made from boron nitride, which has the same chicken-wire❶ lattice structure as graphene. They've made single-layer metal oxides that may serve as highly active catalysts for controlling particular chemical reactions. And they've even trapped water molecules in thin sheets, although what this will be useful for isn't yet clear.

But for now, most of the buzz among flatlanders surrounds just two materials: a compound called molybdenum disulfide (MoS2) and a double layer of phosphorus atoms called phosphorene. Both have tantalizing electronic properties, and the competition between their acolytes is fierce.

Of the two materials, MoS2 had the head start. Originally synthesized in 2008, MoS2 is a member of a broader family of materials called transition metal dichalcogenides (TMDs). The name is just a fancy term for their makeup: one transition metal atom (in this case molybdenum) and a pair of atoms from column 16 of the periodic table (a family known as the chalcogens), which contains sulfur and selenium, among others. Much to the delight of electronics makers, all TMDs are semiconductors. They aren't quite as thin as graphene (in MoS2, twin sheets of sulfur atoms sandwich a middle layer of molybdenum atoms), but they offer other advantages. In the case of MoS2, one is the speed at which electrons travel through the flat sheets—a property called electron mobility. MoS2's mobility is a decent 100 or so centimeters

---

❶ Chicken-wire is a mesh of wire commonly used to fence poultry livestock in a run or coop. It is made of thin, flexible galvanized steel wire, with hexagonal gaps.

squared per volt second (cm^2/vs). That's well below the 1400 cm^2/vs mobility of crystalline silicon, but it's better than the number for amorphous silicon and many other ultrathin semiconductors being tested for use in futuristic applications such as roll-up displays and other flexible, stretchable electronics. MoS2 also turns out to be fairly easy to make, even in large sheets. And that has helped engineers move quickly to testing it in devices. In 2011, for example, researchers led by Andras Kis of the Swiss Federal Institute of Technology in Lausanne reported in *Nature Nanotechnology* that they had made the first transistors using a single layer of MoS2 just 0.65 nanometers thick. Those devices and their successors turned out to have other exceptional properties that rival those of far more developed silicon-based technology. They boast a large on/off ratio, which makes it easy to differentiate between digital 1s and 0s. Since 2011, Kis' group and others have engineered a host of MoS2-based electronic devices including logic circuits and always-on flash memory devices, both of which are widely used in today's computers.

Beyond that, MoS2 has another desirable property known as a direct bandgap, which enables the material to convert electrons into photons of light—and vice versa. That makes MoS2 a good candidate for use in optical devices, such as light emitters, lasers, photodetectors, and even solar cells. Lee, an expert in growing large-area MoS2 films, notes the material is also abundant, cheap, and nontoxic. "It has a bright future," he says.

Tomanek, however, is among MoS2's detractors, saying "it has been oversold". In particular, Tomanek says he isn't convinced that MoS2's electron mobility will ever be high enough to compete in the crowded electronics marketplace. The reason, he says, lies in the material's very makeup. Electrons traveling through it ricochet off large metal atoms in its structure and slow down. That stumbling block will prove temporary, Lee says. Researchers are already learning to navigate around it by growing slightly thicker multilayers of MoS2 that offer zipping electrons alternative routes to bypass roadblocks.

Its rival, phosphorene, has sparked even more excitement. Also known as black phosphorus, phosphorene is one of three different crystal structures—or allotropes—that pure phosphorus can adopt. The others are white phosphorus, which is used in making fireworks, and red phosphorus, used to make the heads of matches. Phosphorene, which consists of a corrugated pattern of phosphorus atoms that lie in two different planes, was first synthesized only last year. But its properties have already made it a materials-science darling. It has an electron mobility of 600, which some researchers hope to increase even further, and its bandgap—the voltage needed to drive a current through it—is tunable. Electrical engineers can adjust the bandgap simply by varying the number of phosphorene layers they stack one atop another, making it easier to engineer devices with the exact behavior desired. "All this makes black phosphorus a superior material," Tomanek says.

  Researchers have made rapid progress toward incorporating it into devices. On 2 March 2014, Zhang and his colleagues at Fudan University reported online in *Nature Nanotechnology* that they had made phosphorene-based field effect transistors, devices that serve as the heart of computer logic circuitry. Two weeks later, Tomanek and colleagues at Michigan State, together with researchers led by Peter Ye, an electrical engineer at Purdue University in West Lafayette, Indiana, reported online in *ACS Nano* that they, too, had made phosphorene-based transistors, along with simple circuits. Unfortunately, phosphorene is unstable in air. "We see bubbles cover the surface after 24 hours and total device failure in days," says Joon-Seok Kim, a phosphorene device maker at UT Austin. The culprit, Lee says, is water vapor, which reacts with the phosphorus, eroding it by converting it to phosphoric acid. Even so, Kim's group at Texas and others are making progress in protecting it. Kim reported at the March meeting of the American Physical Society (APS) in San Antonio, Texas, that he and his colleagues were able to stabilize phosphorene-based transistors for 3 months, counting by encapsulating them in a protective layer of aluminum oxide and Teflon. At the same meeting, researchers from Northwestern University in Evanston, Illinois, reported that a similar strategy gave them stable devices out to 5 months and counting.

  But Lee, for one, is not convinced the fixes will lead to long-term stability. "You can put a capping layer on top, but it just reduces the degradation rate," Lee says. Phosphorene, he argues, is gaining attention because it's easy for researchers to get their hands on: It can simply be peeled off a chunk of black phosphorus with sticky tape, like graphene. "It's a kind of fashion," Lee says. "But that doesn't mean it will have a future."

In the end, there may be plenty of room for both materials. "We're still just at the beginning," says Luis Balicas, a physicist at Florida State University and the National High Magnetic Field Laboratory in Tallahassee. He suggests that over time engineers may wind up favoring MoS_2's strong interactions with light to make solar cells, light emitters, and other optical devices, while harnessing phosphorene's higher electron mobility for making electronic devices. Two-dimensional materials also offer another tantalizing option: They can be stacked like cards in a deck to create the different electronic layers needed in functional electronic devices.

In devices made using conventional 3D materials, neighboring crystalline layers usually bind tightly to one another. But if the atomic lattice of adjacent layers differs by more than 15% or so, the strain at the interface causes one or both layers to crack, a potential device killer. That means electrical engineers must either severely limit their selection of neighboring materials so that the layers can join without strain, or resort to complex workarounds, such as adding "buffer" layers at each interface. With stacked 2D materials, "we don't need to worry about this," Lee says, because they don't form tight bonds with the layers above and below.

That advantage has prompted scientists to build such devices, called van der Waals hetero structures after the weak bonds between adjacent layers. The first ones are already emerging. Last year, Ye's group at Purdue reported that they had used both MoS_2 and phosphorene to make ultrathin photovoltaics (PVs). At the APS meeting, Balicas's group reported similar PVs made by combining layers of TMDs, boron nitride, and graphene. And in February, Geim and colleagues reported online in *Nature Materials* that they had assembled multiple 2D materials to make efficient, thin light-emitting diodes. Such progress has the community of device makers salivating over what may soon be possible. The latest revolution in electronics and optics is just getting started.

(Excerpt from *Science*, May 1, 2015)

Exercises

I. Translate the following technical terms into English.

1. 石墨烯 _____
2. 光学性能 _____
3. 高效催化剂 _____
4. 二硫化钼 _____
5. 分子束外延 _____
6. 非晶硅 _____
7. 逻辑电路 _____
8. 光子 _____
9. 电子迁移率 _____
10. 六角形晶格 _____
11. 超薄半导体 _____
12. 二维材料 _____
13. 闪存设备 _____
14. 电学性能 _____
15. 弱键 _____
16. 凝聚态物理学家 _____
17. 同素异形体 _____
18. 光电探测器 _____
19. 纳米 _____
20. 磷原子 _____

II. Translate the following paragraphs into Chinese.

Yet graphene opened researchers' eyes to a new world of flatland electronics. They saw that similar materials might have novel optical and electrical properties. And because 2D sheets are so thin and mostly transparent, they offered the prospect of creating flexible and transparent electronics that could produce see-through displays of the sort dreamed up years ago by Hollywood. Since then, researchers have been surveying that landscape for richer treasures.

By looking for materials that naturally form 2D sheets and finding ways of stabilizing sheets of atoms that normally want to form a 3D architecture, materials scientists have already come up with dozens of new 2D materials, and many more are likely to follow. They've engineered single-layer silicon (known as silicene), single-layer germanium (germanene), and single-layer tin (stanene). They've created an insulator made from boron nitride, which has the same chicken-wire lattice structure as graphene. They've made single-layer metal oxides that may serve as highly active catalysts for controlling particular chemical reactions. And they've even trapped water molecules in thin sheets, although what this will be useful for isn't yet clear.

Unit 7
Material Science

Text C

Using All Energy in a Battery

Controlled electrode structure improves energy utilization.

By Nancy J. Duneand and Juchuan Li

It is not easy to pull all the energy from a battery. For a battery to discharge, electrons and ions have to reach the same place in the active electrode material at the same moment. To reach the entire volume of the battery and maximize energy use, internal pathways for both electrons and ions must be low-resistance and continuous, connecting all regions of the battery electrode. Traditional batteries consist of a randomly distributed mixture of conductive phases within the active battery material. In these materials, bottlenecks and poor contacts may impede effective access to parts of the battery. Kirshenbaum et al. explore a different approach, in which silver electronic pathways form on internal surfaces as the battery is discharged. The electronic pathways are well distributed throughout the electrode, improving battery performance.

Commercial battery electrodes (anodes and cathodes) are typically created by casting a porous powder composite (a mixture of the active material, a small amount of polymer binder, and a conductive additive such as carbon) onto a metal foil current collector. Electrons are conducted via chains of particles through the composite to the current collector. In contrast, ions move through the liquid or solid electrolyte that fills the pores of the composite. Optimization of both pathways is critical for battery performance. Although this slurry-cast electrode structure works very well, better control of the three-dimensional (3D) architecture would enhance the energy per unit mass and volume of the electrode. Cobb and Blanco recently reported an important step in this direction by creating a cathode consisting of alternating low- and high-density stripes. The low-density stripes provide higher porosity and better access for ions traveling through the liquid electrolyte into the cathode.

Kirshenbaum now reports the fabrication and performance of a silver vanadium phosphate cathode with a well-defined 3D architecture. The cathode consists of relatively dense thick pellets without binder or conductive additives. When Li^+ ions and electrons move into the silver vanadium phosphate particles, $V4^+$ is reduced to $V3^+$ and Ag^+ is reduced to metallic silver; the latter remains at the surface of the active material particles as small silver particles, presumably electrically connected by a thin layer of Ag. Under the right conditions, the silver forms an effective electronic path throughout the electrode, enhancing the insertion of Li^+ into the cathode lattice and hence increasing the amount of accessible energy in the battery.

Such reduction displacement reactions, also known as conversion reactions, occur upon Li^+ reaction with a wide range of binary and bimetallic oxide, fluoride, and sulfide compounds. These materials have potentially very high energy densities that may yield rechargeable and low-cost battery materials. The biggest challenge for practical use of such reversible conversion electrodes is the voltage penalty, where the voltage of the battery during discharge (conversion) is much less than the voltage needed for recharge of the battery (reconversion). It remains unclear how much of this voltage penalty is intrinsic and how much of it is a result of kinetic limitations that could in principle be minimized.

Many studies have shown the advantages of intimate "wiring" of the active battery electrode particles with the electronic conducting component. Robust physical or chemical bonds formed during cosynthesis or annealing of the active particles with conductive fibers give superior performance during charge-discharge cycling compared with traditional electrodes. However, fibrous or templated 3D structures are generally difficult to form as a dense body. Conversion electrodes such as those reported by Kirshenbaum et al. provide improved density by forming an internal conductive network through electrochemical reaction. Bonding of the conversion particles to a conductive carbon fiber scaffold through high-temperature processing not only prevents capacity loss but also reduces the voltage penalty to recharge the battery.

Kirshenbaum analyzes the distribution of discharge products for their model cathode materials in an extraordinary level of detail. They use energy dispersive X-ray diffraction (EDXRD) to probe the intact battery. Although not noted by the authors, the results indicate a slight accumulation of metallic Ag at both faces of the electrode.

This distribution is similar to that in at least one other study of a thick electrode, where the distribution was attributed to the relative transport rates of reactants from opposing directions.

Kirshenbaum focuses instead on the equally intriguing result that the silver metal distribution and discharge performance is very sensitive to the discharge rate: When the current drawn early in the discharge is reduced by a factor of 3, the metallic silver is distributed more uniformly and capacity utilization is higher. The authors attribute this sensitivity to the tendency for the active particles to crack upon rapid lithium addition, which creates more internal surface and nucleation❶ sites for the silver clusters. Support for this idea comes from Woodford et al., who have predicted a critical discharge rate based on particle size and the diffusion-induced stress above which brittle battery particles are likely to fracture.

Other advanced in situ❷ methods are also helping researchers to visualize the complex chemical and structural changes during charge-discharge reactions. Transmission neutron diffraction analysis has revealed strong changes in the active electrode material located close to the edges of a large battery electrode sheet after substantial cycling. This method cannot resolve any gradients across the thickness of the battery electrodes, but can map lateral inhomogeneities along the battery area from edge to edge. Synchrotron radiation X-ray tomographic microscopy provides dramatic maps of changes in both electrode structure and chemical content with cycling. Ebner recently used this method to study Li addition into SnO particles in a carbon matrix, revealing swelling and crack formation in 20-μm particles. Yang used transmission X-ray microscopy combined with X-ray absorption near-edge structure to reveal much more subtle changes in a cycled intercalation electrode, reporting not only a distortion in the particle shape, but also a redistribution of Mn, Ni, and Co transition metals and formation of a new phase.

Kirshenbaum's study is an exciting step toward understanding how optimized

---

❶ Nucleation is the first step in the formation of either a new thermodynamic phase or a new structure via self-assembly or self-organization.

❷ In situ is a Latin phrase which means on site, or in position. In chemistry, in situ describes the way a measurement is taken, that is, in the same place the phenomenon is occurring without isolating it from other systems or altering the original conditions of the test.

battery electrode architectures can maximize the energy per unit volume and weight. Silver compounds may be too expensive for applications other than medical ones, but bimetallic polyanionic materials containing Cu or Fe have promise as active electrode materials with widespread application. To further improve access to full capacity, future, thicker electrodes could also include gradients in morphology spanning the thickness of the electrode and the distance from the electrode terminal. Using the battery chemistry itself to drive the formation of the electrode structure is an elegant approach toward such an optimized structure.

(Excerpt from Science, January 9, 2015)

Exercises

Reading Comprehension

Directions: Answer the following questions based on the information from the text.

1. What are the conditions for a battery to discharge?
2. Why is it difficult to pull all the energy in a traditional battery?
3. How did Kirshenbaum et al. improve the battery performance?
4. Describe the working principles of a commercial battery and the improvements made by Cobb and Blanco.
5. What are the methods that help researchers to visualize the complex chemical and structural changes during charge-discharge reactions?

Unit 8
Engineering

导读

本单元的文章均涉及工程领域里新技术的研发和应用。Text A介绍了美国国家航空航天局星座计划的波折和争议，阐述了该计划中重型载人运载火箭测试中心和建造中心所采用的新技术和新方法，并对该火箭的发射方式和意义进行了详细说明。Text B涉及工程领域中柔性设计的新概念。作者阐述了柔性设计概念的起因、内容和优点，并详细介绍了其团队根据柔性设计理念设计并制造一体化雨刮器和飞机机翼等部件的过程。Text C介绍了污水处理工程方面的新技术和人们对使用净化后的污水所持有的不同态度，具体阐释了两种使用净化后污水的方法，并以加州圣迭戈先进净水设施为例说明这种新技术的应用及意义。

Part I Intensive Reading

Text A

Birth of a Rocket

Is NASA's Space Launch System a flying piece of congressional pork[1] or our best shot at getting humans to deeper space?

By David H. Freedman

It is the sad tale of NASA's Michoud Assembly Facility, the sprawling New Orleans complex where the space agency had for decades built its biggest rockets. After the space shuttle's last flight in 2011, Michoud's massive hangar-like facilities were rented out to Hollywood studios.

---

[1] Congressional pork is part of a large political movement known as "pork barrel politics" in America. Pork barrel is a metaphor for the appropriation of government spending for localized projects and is secured solely or primarily to bring money to a representative's district.

But lately a growing cadre of NASA engineers and other workers have been engaged on an important new production here. Michoud is back in the rocket-making business, serving as a factory for the biggest, most ambitious space vehicle ever to undergo construction: the Space Launch System, often called by its acronym, SLS.

The SLS is the rocket in which NASA hopes to thunder a crew of astronauts skyward from Cape Canaveral, Fla., for roughly a year's journey to the surface of Mars while hauling the living quarters, vehicles and supplies they will need to spend at least a few weeks shuffling through the rusty dust there. That mission is still about 25 years away. But between now and then, the SLS could carry people to Earth's moon and an asteroid and send a probe to search for life on Europa, one of Jupiter's moons. It is an interplanetarily groundbreaking project, one of the most audacious NASA has ever undertaken.

Why, then, do so many people seem to hate it?

● Replacing the Shuttle

After the Giddy Triumph of the Apollo moon exploration program in the 1960s and early 1970s, the space shuttle was supposed to make Earth-orbit access relatively cheap and routine. Instead the shuttle averaged more than $1 billion a trip, flew only a few times a year and was twice afflicted by catastrophe. In 2004, a year after *the Columbia* disintegrated on reentry, killing seven people, President George W. Bush charged NASA with replacing the shuttle with a more Apollo-like program that would bring us back to the moon and then to Mars. The resulting effort, called Constellation, led to the design of two new Ares rockets, a crew launch vehicle and a giant, Saturn V-like version intended to haul cargo. But by 2011, after having burned through some $9 billion, all Constellation had produced was an *Orion* crew capsule that was being constructed by Lockheed Martin and a rocket that had been launched once as a test. President Barack Obama canceled the program, directing NASA to refocus its energy on a mission to an asteroid. The agency was to turn to the private sector for an orbital ferry service to get cargo and crew to the International Space Station (ISS).

Still, many in Congress pushed hard to continue the quest for a new heavy-lift rocket capable of getting humans to the moon and Mars. The resulting compromise was the SLS, a single big rocket for both crew and cargo that would eschew much of the new technology planned for Ares and instead rely on space shuttle engines, boosters and tanks for most of its kick. The SLS was Ares on the cheap.

From the beginning, the SLS has been dogged by the perception that Congress cooked it up to protect jobs at NASA and its major contractors. Some critics deride the SLS as the "Pork Rocket" or "Senate Launch System". Southern senators whose states are home to large NASA or contractor facilities have indeed been the SLS's loudest proponents in Congress.

And a big program—and rocket—it is. The SLS will initially have a bottom core stage powered by four RS-25 space shuttle engines that use standard liquid hydrogen and oxygen fuel. Attached to each side of the core stage will be solid rocket boosters, which provide the extra push needed to get the heavy rocket airborne. A second stage, atop the first, will take over at an altitude of about 50 kilometers to push the rocket into orbit, and the Orion crew capsule will sit on top of the entire structure. At 98 meters, the rocket will be slightly shorter but more powerful than a Saturn V, which powered every manned mission to the moon, and will carry three times the payload of the shuttle. None of the components are designed to be reusable. Over the next decade, SLS upgrades will include more powerful engines and boosters. The eventual Mars-capable SLS would get even more power in its upper stage, giving it twice the thrust of the first version.

Critics charge that by specifying that the SLS rely on shuttle components, Congress ensured that the shuttle's big aerospace contractors would profit. Others contend that the shuttle recycling approach will leave the SLS a troubled Frankenrocket with stitched-together parts from a dead program. Estimates of the SLS's final cost vary wildly. NASA has publicly projected that it will take $18 billion to get the SLS to first launch. But a leaked internal study came up with a cost of more than $60 billion over the next 10 years. Others predict that delivering a crew to Mars will cost up to $1 trillion.

Critics insist that the government and public will never back their enthusiasm for space exploration with the many hundreds of billions of dollars the SLS's grandest missions will require. Several analyses have suggested that we can get to deep space and Mars without a heavy-lift rocket. It might be cheaper, some argue, to rely on smaller rockets to heft into low-Earth orbit the fuel, components and materials needed to construct deep-space vehicles and then build the big craft there. The "SLS is only adding small incremental improvements to technology developed 40 years ago," says James Pura, president of the Space Frontier Foundation, an advocacy group dedicated to advancing space exploration.

Despite these objections, SLS mission planning is underway. A 2018 first flight will send a crewless SLS and *Orion* out well past the moon, and a second, not yet formally scheduled flight will do much the same with a crew perhaps a few years later, taking humans farther from Earth than ever before. Right now a crewed asteroid visit is tentatively planned for the mid-2020s, with a human mission to Mars to follow in the 2030s.

● The Rocket Factory

NASA tests its biggest rockets at Stennis Space Center, which lies in a web of lakes, rivers, bayous and canals near the southernmost tip of Mississippi. There are three reasons for the center's proximity to water: The activities at Stennis require access to large barges, to marine construction expertise and to a ready way to cool giant slabs of metal exposed to temperatures approaching those found on the surface of the sun.

Each test stand here is a huge metal-and-concrete structure. We climb up through one of the stands, and along the way I am shown a control room that would not look out of place in a circa 1950s Soviet power plant—mostly steam gauges and big, clunky dials. From the top of the stand, however, I can see that Stennis is actually awash in upgrades. Canals and roads are being reworked to handle larger loads, and the test stands themselves are getting renovations and reinforcements because the SLS is going to subject them to greater stresses than any previous rocket. Throughout an approximately nine-minute test-firing, thousands of nozzles will shoot high-pressure jets of water at the stand's walls—not for cooling but to tamp down ferocious vibrations that could otherwise rip the stand apart. Even before the SLS, no private structure was allowed within 13 kilometers of the stands because the sound waves alone from a test could shake it apart. And the SLS engines will generate the most powerful rocket thrust ever produced on Earth.

Just across the Mississippi-Louisiana border, a few hours away via canal, sits Michoud, which I visit the next day. In contrast to the isolation of Stennis, Michoud is in the middle of an industrial area on the outskirts of New Orleans. In some ways, Michoud is a factory like any other, with welding stations, forklifts, cranes and parts bins. It is just all done on a much larger scale.

Inside, Michoud is gleaming. To tour the complex is to watch it fill up, minute by minute, with new gear—towering robot arms that can move at blinding speed, wheeled platforms and cranelike handlers that whisk components weighing tens of metric tons from one station to another, parts-organizing systems that ensure that an engine consisting of hundreds of thousands of parts does not end up with one too many or few. When you build a machine as powerful as an SLS rocket engine, you must have a very low tolerance for assembly deviation. "If our parts-tracking system told us that one of these tiny washers here is left over, all work would stop until we found it," says Patrick Whipps, one of NASA's managers at Michoud.

Many of the components that will go into the rockets built here originated in other vehicles. Yet new manufacturing equipment and methods should make those components much less expensive to build than they have been in the past, Whipps adds. Upgrades include a friction-stir welding machine the size of a municipal water tower tank. Massive aluminum-alloy rocket sections can be dropped whole into this leviathan, where drills will meld the two sections together. It is the largest machine of its type in the world.

The SLS goes beyond shuttle technology in many other ways as well. To analyze the stresses on the SLS from buffeting and other aerodynamic instabilities during its climb through the atmosphere, NASA turned to state-of-the-art fluid dynamics software. In addition, new avionics and digital controllers relying on computer chips that are several generations ahead of those used in the space shuttle will enable automated flight and engine controls to react many times faster to sudden changes and dangerous conditions

Leftover shuttle engines will get the SLS airborne for the first four flights, but new versions will be needed starting in the 2020s. For those, NASA is using machines that will produce the thousands of required coin-sized turbine blades by laser-welding powdered metal into the right shapes instead of individually machining them, cutting production time for an engine's worth of blades from a year to a single month. "We're using computer control everywhere to minimize labor costs and improve precision," Gerstenmaier says.

- **The Case for SLS**

When the SLS program is in full swing, the aim will be to turn out at least two rockets a year. In the rocket world, that is mass production. But it will grind to a halt if NASA cannot convince the American public that the SLS is worth building.

The two broadest objections—that $18 billion is too much to spend on a rocket and that we should focus on sending probes and robots, not humans, into space to do science—can be addressed as matters of perspective. Eighteen billion dollars is not all that much for the capability of sending humans to another planet and back; it cost a third more than that to improve traffic flow in Boston via the "Big Dig❶". It is easy to claim there are cheaper ways of doing it, but NASA's success and safety records have set the bar high, and it is unlikely that the American public would put up with higher chances of a catastrophic failure in order to shave off what amounts to a few thousandths of the federal budget. As for sticking with probes and robots, the case is often made that the science haul from a human-crewed mission is likely to be bigger than what a probe or rover can deliver. But the real justification for human spaceflight is to take steps toward expanding the human race's stomping grounds.

The SLS does have many fans. These supporters include NASA's current

---

❶ The Central Artery/Tunnel Project, known unofficially as the Big Dig, was a megaproject in Boston that rerouted the Central Artery of Interstate 93, the chief highway through the heart of the city, into the 3.5-mile Thomas P. O'Neill Jr. Tunnel. The Big Dig was the most expensive highway project in the U.S., and was plagued by cost overruns, delays, leaks, design flaws, charges of poor execution and use of substandard materials, criminal arrests, and one death. The project was originally scheduled to be completed in 1998 at an estimated cost of $2.8 billion (in 1982 dollars). However, it was completed at a cost of over $14.6 billion ($8.08 billion in 1982 dollars) as of 2006.

leadership and rank and file, a number of space experts and a growing chunk of the American public, much of which was thrilled last December by the flawless orbital flight of the *Orion* crew capsule that will be sitting atop the SLS when it heads into deep space. The experts among them can easily argue, point by point, with the critics.

Use smaller rockets to heft components and fuel into space for orbital assembly? Some 500 metric tons of materiel will be needed for a crewed Mars mission, Gerstenmaier calculates. That is a feat that the SLS could manage in four launches but that would take at least two dozen launches of a maxed-out Delta IV. Gerstenmaier contends that every one of those launches raises program risk a bit because the worst things are most likely to happen in the first minute of a mission.

But the most significant potential drawback to a lift-it-in-small-chunks approach, Gerstenmaier says, is the massive amount of in-orbit construction that would be required, including habitats, interplanetary vehicles and fuel depots. "You'd have a huge number of dockings; you'd be fabricating in space," he says. "Inevitably some of the pieces wouldn't work right and would be difficult to fix there. It adds an enormous amount of complexity and risk." The SLS's sheer girth will also allow packing in bulkier, ungainly payload shapes up to 10 meters across, such as those with solar panel and antenna arrays, that otherwise would have to be complexly folded and thus more vulnerable to damage or malfunction.

Another big advantage to the heavy-lift route: Some of an outsize rocket's extra thrust can be converted into higher speeds that get spacecraft to their destinations more quickly. That is a critical consideration for crewed flight to Mars, where radiation exposure and supply requirements set tight upper limits on mission duration. Distant robotic missions benefit, too, because planning for follow-up missions has to wait for data to come in from predecessors to maximize the scientific returns. Because of its sheer power, the SLS can send missions into deep space using its own fuel, as opposed to gravitationally slingshotting around planets as the Voyager and Galileo missions did.

The "SLS will cut the time for a Europa visit from six-plus years to 2.5 years," says Scott Hubbard, a consulting professor of aeronautics and astronautics at Stanford University. "It would be an enabler for a very compelling scientific mission."

Add these shorter transit times to the higher payload masses and packaging flexibility, and you have a powerful case for a heavy-lift rocket.

The same goes for SpaceX. Yet new space is not as natural a source for deep-space rockets as it is for transport to the ISS and back. There is no existing market, and none envisioned, for deep-space exploration beyond the handful of missions NASA has tentatively planned for the SLS. That eliminates the opportunity for SpaceX to leverage development costs for a heavy-lift rocket over various commercial customers, as it has with its smaller rockets.

Hewing to the tried and tested instead of innovating might be a recipe for failure in the automobile, cell phone or software industries, but when it comes to zipping a crew of heroes into deep space on the wings of a barely controlled explosion, a certain level of conservatism is not necessarily a bad thing. SpaceX suffered several explosions and losses of control in its earlier rockets—par for the course in the development of new designs. Last October a crew member was killed in the explosion of a prototype rocket that Virgin Galactic built to bring tourists into suborbital space—just three days after the explosion of a crewless rocket built by private company Orbital Sciences, one that was headed to the ISS.

These accidents serve as reminders that in spite of decades of experience, rocketry is hard. It carries a high risk of pure catastrophe. That is one reason leaders at the Inspiration Mars Foundation, a privately funded organization that has been trying to facilitate a mission to Mars, are among those who have, after initial skepticism, been lining up behind the SLS. Other Mars experts agree.

- **Escape Velocity**

For 500 seconds on a cool night this past January, one of the Stennis Space Center's hulking engine tests turned into a fireball. It was the first test of an R-25 shuttle engine since 2009, and it went perfectly. If the successful tests keep coming, time may be on the SLS's side. The longer the program lasts—if it remains on budget and on time—the more it will stand as its own proof of concept. In its first three years, the program has achieved smooth and rapid progress, gliding through design reviews and entering into early manufacturing steps. That is blindingly fast for a major new human-rated rocket. Only a few glitches have cropped up; those insulation gaps were just about the worst of them, and the problem was quickly fixed with a layer of adhesive.

Anything could happen in the years ahead, under new presidents and congresses, contends Joan Johnson-Freese, a professor at the U.S. Naval War College who specializes in space. Maybe the consensus in government will become that we should abandon Mars for now and focus on setting up a base a little closer to home. "Some in Washington have an almost criminal nostalgia for the moon," she says. Others think NASA should forget both the moon and Mars for now and concentrate on asteroids, not only because they may contain answers to important questions about the origins of the solar system but also because we might learn how to divert or destroy any that end up heading toward Earth.

But the allure of Mars remains widespread. Lately that allure has been building, as it dawns on more people that we could reach the Red Planet within their lifetime. At the moment, there are no showstoppers in sight for the SLS. That claim alone, which cannot be made for any alternative Mars rocket proposal, may ensure that the project stays the course. That should be good enough to make the SLS the rocket that takes us to Mars. And if it does, the criticisms will be quickly forgotten.

(Excerpt from *Scientific American*, June 2015)

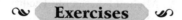

I. Reading Comprehension

• **Section One**

Directions: Answer the following questions based on the information from the text.

1. What is the Space Launch System? Why do many Americans object to it?
2. What is the function of Stennis Space Center? Describe its renovations and reinforcements mentioned in the text briefly.
3. What is the function of Michoud? Describe its new manufacturing equipment and methods mentioned in the text briefly.

4. Summarize the ways of sending the heavy-lift rocket into deep space or to Mars in the future mentioned in the text and their advantages and disadvantages.
5. What is the prospect of the Space Launch System program according to the text?

● Section Two

Directions: Write an abstract based on the text in no more than 200 words.

Abstract:
Key words:

II. Vocabulary

● Section One

Directions: Choose the explanation that is closest in meaning to the underlined part in each sentence.

1. It is the sad tale of NASA's Michoud Assembly Facility, the sprawling New Orleans complex where the space agency had for decades built its biggest rockets.
    A. a combination of emotions and impulses that have been rejected from awareness but still influence a person's behavior
    B. a whole structure made up of interconnected or related structures
    C. any loosely bonded species formed by the association of two molecules
    D. a group of system of different things that are linked in a complicated way
2. The SLS is the rocket in which NASA hopes to thunder a crew of astronauts skyward, Fla., for roughly a year's journey to the surface of Mars while hauling the living quarters, vehicles and supplies they will need to spend at least a few weeks shuffling through the

Unit 8
Engineering

rusty dust there.

A. moving people or things around so as to occupy different positions

B. walking by dragging one's feet along or without lifting them fully from the ground

C. shifting one's position while sitting

D. sorting a number of things hurriedly

3. It is an interplanetarily groundbreaking project, one of the most <u>audacious</u> NASA has ever undertaken.

   A. loud enough to be heard   B. imprudent

   C. foolish bold   D. daring

4. The resulting compromise was the SLS, a single big rocket for both crew and cargo that would eschew much of the new technology planned for Ares and instead rely on space shuttle engines, boosters and tanks for most of its <u>kick</u>.

   A. strike with the foot

   B. a sudden forceful jolt

   C. strength

   D. boost

5. But when it comes to <u>zipping</u> a crew of heroes into deep space on the wings of a barely controlled explosion, a certain level of conservatism is not necessarily a bad thing.

   A. locking together two toothed edges by means of sliding tab

   B. moving or delivering at high speed

   C. compressing so that it takes less space in storage

   D. enclosing in by locking the door

6. In its first three years, the program has achieved smooth and rapid progress, gliding through design <u>reviews</u> and entering into early manufacturing steps.

   A. summary at the end that repeats the substance of a longer discussion

   B. formal or official assessment or examination of something

   C. essay or article that gives a critical evaluation (as of a book or play)

   D. judicial reexamination of the proceedings of a court (especially by an appellate court)

7. "Some in Washington have an almost <u>criminal</u> nostalgia for the moon," she says.

   A. (of an action or situation) deplorable and shocking

   B. involving or being or having the nature of a crime

C. bringing or deserving severe rebuke or censure

D. of or relating to illegal act or activity

8. That is a critical consideration for crewed flight to Mars, where radiation exposure and supply requirements set <u>tight</u> upper limits on mission duration.

  A. fixed   B. close-fitting   C. tense, irritated   D restricted

9. Add these shorter transit times to the higher <u>payload</u> masses and packaging flexibility, and you have a powerful case for a heavy-lift rocket.

  A. effective load   B. capacity   C. downloading   D. uploading

10. "SLS is only adding small <u>incremental</u> improvements to technology developed 40 years ago," says James Pura, president of the Space Frontier Foundation, an advocacy group dedicated to advancing space exploration.

  A. increasing gradually   B. incredible   C. imaginable   D. limited

● **Section Two**

Directions: There are two or three meanings for each semi-technical term underlined in the following sentences. Choose the correct one according to the context.

1. But between now and then, the SLS could carry people to Earth's moon and an asteroid and send a <u>probe</u> (an unmanned exploratory spacecraft designed to transmit information about its environment; an investigation into a crime or other matter) to search for life on Europa, one of Jupiter's moons.

2. But by 2011, after having burned through some $9 billion, all Constellation had produced was an *Orion* crew <u>capsule</u> (a tiny container for a dose of medicine; a spacecraft designed to transport people and support human life in outer space; shortened version of a written work) that was being constructed by Lockheed Martin and a rocket that had been launched once as a test.

3. Over the next decade, SLS upgrades will include more powerful engines and <u>boosters</u> (the first stage of a multistage rocket; someone who is an active supporter and advocate).

4. When you build a machine as powerful as an SLS rocket engine, you must have a very low <u>tolerance</u> (diminution in the body's response to a drug after continued use; an allowable amount of variation of a specified quantity, especially in the dimensions of a machine or part; the power of capacity of an organism to tolerate unfavorable

environmental conditions) for assembly deviation.

5. "If our parts-tracking system told us that one of these tiny washers (someone who washes things for a living; seal consisting of a flat disk placed to prevent leakage; a home appliance for washing clothes and linens automatically) here is left over, all work would stop until we found it."

6. Massive aluminum-alloy rocket sections can be dropped whole into this leviathan, where drills (a tool with a pointed end or cutting edges for making holes in hard substance; systematic training by multiple repetitions) will meld the two sections together.

7. NASA is using machines that will produce the thousands of required coin-sized turbine blades by laser-welding powdered metal into the right shapes instead of individually machining them, cutting production time for an engine's worth (high value or merit; an amount of a commodity equivalent to a specified sum of money; the amount that could be achieved or produced in a specified time) of blades from a year to a single month.

8. "You'd have a huge number of dockings (enclosed areas of water in a port for the loading, unloading, and repair of ships; facilities for a spacecraft to join with a space station or another spacecraft in space); you'd be fabricating in space," he says.

9. Only a few glitches have cropped up; those insulation gaps were just about the worst of them, and the problem was quickly fixed with a layer of adhesive (a substance used for sticking objects or materials together; an abnormal union of surfaces due to inflammation or injury).

10. And the SLS engines will generate the most powerful rocket thrust (a forceful attack or effort; the propulsive force of a jet or rocket engine; a principal purpose or theme of a course of action or line of reasoning) ever produced on Earth.

● **Section Three**

Directions: Match the Chinese terms with their English equivalents.

1. 猎户宇航员舱   A. reentry
2. 小行星   B. heavy-lift rocket
3. 固体火箭助推器   C. asteroid
4. 近地轨道   D. low-Earth orbit
5. 涡轮叶片   E. thrust

6. 标准液态氢和液态氧燃料
7. 推力
8. 重返大气层
9. 重型运载火箭
10. 气压计
11. 深空探测
12. 轨道运输服务
13. 零件跟踪系统
14. 空气动力的不稳定性
15. 装配偏差

F. turbine blade
G. steam gauge
H. standard liquid hydrogen and oxygen fuel
I. deep-space exploration
J. aerodynamic instability
K. *Orion* crew capsule
L. assembly deviation
M. parts-tracking system
N. solid rocket booster
O. orbital ferry service

III. Questions for Discussion

Directions: Work in groups and discuss the following questions.

1. What does the controversy on the Space Launch System imply?
2. What is the ultimate aim of exploiting the Space Launch System for NASA?
3. What is the latest spacecraft development in China?

Unit 8
Engineering

Part II   Extensive Reading

Text B

Shape-Shifting Things to Come

Flexible, one-piece machines could soon make today's assemblages of rigid parts look like antiques.

By Sridhar Kota

As I sat behind the wheel, it struck me that my windshield wiper was a ludicrous waste of engineering effort. The wiper frame, which holds the disposable blade, has to be highly flexible. It must keep the blade pressed against the glass as it moves back and forth across a variable contoured surface. Moreover, it must be able to do so on a number of car models, each of which has its own windshield geometry. Our response to this need for flexibility? A complicated system of rigid bars, links and pivots.

At the time I had another burgeoning interest—elastic, or compliant, design, which involves building flexible, strong machines from as few pieces as possible. My colleagues and I had already succeeded in building machines from a single piece of material. For instance, in 1993 my graduate students G. K. Ananthasuresh, Laxman Saggere and I built a no-assembly compliant stapler. But the windshield wiper struck me as a perfect test case. A one-piece, or monoform, wiper would virtually eliminate assembly. If successful, such a project would be more than an exercise in engineering minimalism. Most of the cost of manufacturing a windshield wiper goes into its assembly. It should surprise no one that the production of such assembly-intensive products moved offshore to low-wage countries long ago.

My colleagues and I did not get around to designing the one-piece windshield wiper right away. For the past two decades most of my research has focused on general principles for elastic design—developing the theoretical tools that engineers need to design and build compliant devices. But we did eventually design that windshield wiper. In fact, we have used elastic design to build miniature monoform motion amplifiers, flexible airplane wings, robot snakes, and other machines, each one an expression of a new engineering paradigm whose time has come.

● **Living Machines**

We are more familiar with compliant machines than we might think. Perhaps the earliest and most elegant example is an archer's bow. As the archer draws the bow, elastic energy is stored slowly and then released quickly to propel an arrow. This strong, flexible mechanism can be used many times with precision and without failure. A newer example is the cap of a shampoo bottle: It is a monoform device that combines an easy-opening cap and a screw-on sealing collar without a mechanical hinge.

The most successful elastic designs exist in nature. Tree branches, bird wings, crab legs and elephant trunks are all flexible and strong. Their components either grow out of one another or are bonded together with strong, self-regenerating interfaces. Unlike systems of gears, sliders and springs, they bend, warp and flex by exploiting their inherent elasticity.

Humans have accumulated millennia of experience designing strong and rigid structures such as bridges and buildings. For the most part, we do this by using materials that are strong and stiff. If the stresses get too high, we simply add more material to share the load or increase its stiffness. Stiffness, in this paradigm, is good; flexibility, bad. Indeed, with rigid structures, deflection—the tendency to deform, or give under stress—is desirable only if you are designing for earthquake resistance. Compliant design, in contrast, embraces deflection. If the stress on a flex point gets too high, we make it thinner, not thicker, because the function of a compliant structure is to exploit elasticity as a mechanical or kinematic function.

In the case of the shampoo bottle cap, the stress is focused on the thin polymer section that connects the lid to the base. Disposable forceps have much the same design. When the stresses are concentrated in a thin, discrete area, the flexion is referred to as lumped compliance. Researchers have been studying lumped compliance since the 1950s. More recently, Ashok Midha of the Missouri University of Science and Technology, Larry Howell of Brigham Young University, Shorya Awtar of the University of Michigan and Martin L. Culpepper of the M.I.T. have all done excellent research on the subject, demonstrating applications of lumped compliance in precision instruments and nanopositioning devices.

The archer's bow, in contrast, has no such localized flexural zone: It displays "distributed compliance" throughout its whole length. Distributed compliance is essential for building flexible machines that have to do heavy work—wings that must keep planes in the air, for example, or motors that must run for millions of cycles. When I began my work in this field, I could find no theoretical underpinnings or general methods for designing machines with distributed compliance. Naturally, that is where I focused my efforts, and it is where my interest remains.

- **Starting Small**

I started working on flexible, one-piece machines not because they seemed like intriguing novelties but because in certain applications, designing for no assembly is a necessity. I began my career studying large mechanical systems such as automotive transmissions. In the early 1990s, however, I found myself designing truly tiny machines—micro electromechanical systems (MEMS). This was largely a circumstance of that era. Telecommunications companies were starting to develop optical switches for fiber-optic networks; they would use minuscule motors to change the angle of mirrors very quickly to route an optical signal in one direction or the other. Not long after I began reading Vogel[1] and exploring elastic design, I embarked on a project with Steven Rodgers and his team at Sandia National Laboratories' microsystems division, where a monoform design seemed perfect.

Sandia needed to build a linear motor with sufficient output displacement to do work—at least 10 microns. Yet the fabrication constraints of electrostatic motors limit their motion to two microns. I knew I could not simply miniaturize, say, a geared transmission. Even if we could find someone with steady enough hands to assemble gears, hinges and shafts with dimensions in the one- to two-micron range, the resulting machine would be too sloppy for modern engineering. At MEMS scale,

---

[1] Steven Vogel was an American biomechanics researcher and a prolific author of popular works on the intersection of physics and biology.

a machine with a tenth of a micron of clearance is about as useful as a Tinkertoy[1]. Besides, MEMS devices are batch-fabricated much the same way as integrated circuits, tens of thousands in an area the size of a thumbnail. Given all that, I designed a monoform motion amplifier to generate 20 microns of output motion when integrated with the electrostatic motor.

By 1998 we had the motor and amplifier humming away. It had been running for more than 10 billion cycles with no end in sight. But to my mind, the most impressive thing was that the entire motion amplifier, with all its complexity and flexibility, consisted of a single piece of polysilicon.

● **Flexible Fliers**

Of all of the reasons that I have chosen to study compliant design, the one I find most compelling is shape adaptation, or "morphing". The ability to alter the geometry of a structure in real time enables nature's machines to operate with the utmost efficiency. Compare this adaptability with the fixed geometries of the engineered world—automotive drivetrains, airplane wings, engines, compressors, fans, and so on. These and practically all other conventionally designed machines are most efficient under very specific conditions. They operate suboptimally the rest of the time. An aircraft, for example, experiences a variety of flight conditions as it goes from point A to point B—changing altitude, speed, even weight as its fuel is consumed—which means that it is almost constantly operating less efficiently than it could. Birds, on the other hand, can take off, land, hover and dive by effortlessly adjusting the configuration or shape of their wings on demand.

Back in the mid-1990s, I wondered if anyone had ever attempted to change a wing's shape (camber) during flight to improve performance. I was amazed to discover that the Wright brothers had pioneered a different type of wing morphing—wing twist—in their original flier. After a few months of study, I came across a small blurb in a newspaper about flexible-wing research that was conducted in the late 1980s at Wright-Patterson Air Force Base in Ohio. The engineers there called their

---

[1] Tinkertoy is the simple name for Tinkertoy Construction Set, a toy construction set for children. It was created in 1914 in the U.S. with a purpose of inspiring children to use their imaginations.

goal a mission-adaptive wing (MAW). I understood that a morphing wing was not a wacky idea, so I contacted the researchers to ask whether they might be interested in reviewing my design. Their reaction was overwhelming.

They explained that most, if not all, past attempts to create a morphing wing have employed rigid structures—complex, heavy mechanisms with scores of powerful actuators to make a wing structure flex to different geometries. Their adaptive wing showed aerodynamic promise, but the structure was deemed too heavy and complex for practical application.

This did not surprise me. Designing a practical variable-geometry wing would involve satisfying many conflicting requirements. The wing must be lightweight, strong enough to withstand thousands of kilograms of air loads, reliable enough to operate for hundreds of thousands of hours, easy to manufacture and maintain, and durable enough to withstand chemical exposure, ultraviolet radiation and significant temperature changes. The conceptual and software tools in use at the time were never intended to design monoform machines, let alone ones that satisfied so many competing demands

The flexible-wing design I submitted to Wright-Patterson exploited the elasticity of the test components, which were completely conventional aerospace-grade materials. The wing had an internal structure designed to deform easily when a compact internal motor applied force, and it still remained stiff when powerful forces were exerted externally in the wind-tunnel test. The senior engineers at Wright-Patterson were excited about the design, and so was I. In fact, I was so enthusiastic that in December 2000 I founded a company, FlexSys, to develop practical applications of compliant design.

Six years later, after much development and several successful wind-tunnel tests, we managed to get a prototype of the flexible wing affixed to the underside of a Scaled Composites White Knight aircraft for flight tests in the Mojave Desert. The wing was mounted below the jet's body and fully instrumented to measure lift and drag. Its coefficient of lift varied from 0.1 to 1.1 without increasing drag; that translates to a fuel-efficiency boost of up to 12 percent in a wing designed to take full advantage of the new flexible flap. Considering that U.S. airlines consume about

16 billion gallons of jet fuel every year, these seemingly small percentages could be significant. The wing was also simpler, with no moving parts in the morphing mechanism. As a result, it would be more reliable and have a better weight-to-power ratio.

The real test for shape-adaptive aircraft wings will come when flexible-control surfaces completely replace conventional flaps. Working with U.S. Air Force research labs, FlexSys designed and built a continuous surface that bends (cambers) and twists spanwise to maximize aerodynamic performance in place of drag-producing trailing-edge flaps. We have retrofitted a Gulfstream Aerospace GIII business jet with our FlexFoil variable-geometry-control surfaces instead of conventional flaps. In addition to significant fuel savings, our design is expected to reduce aircraft noise: According to NASA, much of the noise involved in landing a plane is caused by vortices generated at the sharp edges and gaps between the deployed trailing-edge flaps and the fixed parts of the wing. We have included transition surfaces to eliminate these gaps. Flight tests at NASA's Neil A. Armstrong Flight Research Center are scheduled to take place in July.

- **Creepers and Crawlers**

Soft-bodied animals such as worms and octopuses lack any apparent skeletal structure, and yet they can move vigorously and gracefully. For the most part, they accomplish this through what is called elastofluidics. In engineering terms, their bodies are hydrostats—they consist of an arrangement of connective tissue fibers and muscles surrounding a pressurized, liquid-filled cavity. A study of the anatomy of these creatures commonly reveals a cross-helical arrangement of fibers and muscles surrounding the internal organs, which occupy the liquid-filled core. The cross-helical fibers serve as antagonists against the fluid pressure generated by muscle contraction; the orientation of the fibers determines the range of motion. Many variants of hydrostatic skeletons exist throughout the animal world. The arms of an octopus are muscular hydrostats. An elephant's trunk employs tightly packed muscle fibers around a hydrostatic body.

Our research on elastofluidics is still in its infancy, but our hypothesis is that these elements could serve as components for constructing "soft robots" and other devices that can safely interact with humans and the environment. The earliest applications, however, will most likely be in the field of orthotics. For instance, patients suffering from arm contracture caused by muscle hardening, joint deformity or joint rigidity could use a flexible orthotic device that gently forces their arm back into functional position for daily activities.

- **Compliance Is Appreciated**

The basic research we started in 1992 has resulted in a trove of useful insights and systematic design methods. Some of the devices we have developed over the years are nearing commercialization. We have completed weather testing and finished the production mold for our monoform windshield wiper frame, and discussions are under way with automakers and suppliers for implementing it as a rear wiper. The monoform wiper is made of glass-filled thermoplastic polymer and works properly in both frigid and hot conditions. It will not snap or twist even when breaking loose ice and snow. When it comes to market, it should be much more durable and reliable and cheaper to manufacture than any competing device.

Our flexible aircraft wings are technically ready for commercial implementation right now. Replacing the outer 15 percent of an existing flap with a variable-geometry subflap for cruise trim alone could save 5 percent in jet fuel. Replacing the entire flap with a seamless FlexFoil offers about 12 percent fuel savings on new designs. Once the industry gains confidence in flexible wings, we believe it is likely that they will replace hinged flaps completely in future fixed-wing aircraft of all types.

Cases abound in the automotive, appliance, medical and consumer sectors where elastic design could drastically reduce the number of parts used in any given device. The biggest challenge is getting the word out to industrial designers. Another challenge remains: There are currently no easy-to-use software tools available for exploring elastic design. With a contract from the National Science Foundation, FlexSys is developing software along these lines.

(Excerpt from *Scientific American*, May 2014)

Exercises

I. Translate the following technical terms into English.

1. 柔性设计 _____
2. 雨刮器 _____
3. 工程范式 _____
4. 无缝弹性机翼 _____
5. 蛇形机器人 _____
6. 刚性结构 _____
7. 光纤网络 _____
8. 位移输出 _____
9. 水压调节器 _____
10. 形状渐变 _____
11. 气动力负荷 _____
12. 伸展性 _____
13. 静电马达 _____
14. 铰链 _____
15. 耐候测试 _____
16. 痉挛 _____
17. 集中柔度 _____
18. 分布柔度 _____
19. 齿轮传动 _____
20. 风洞测试 _____

II. Translate the following paragraphs into Chinese.

I started working on flexible, one-piece machines not because they seemed like intriguing novelties but because in certain applications, designing for no assembly is a necessity. I began my career studying large mechanical systems such as automotive transmissions. In the early 1990s, however, I found myself designing truly tiny machines—micro electromechanical systems (MEMS). This was largely a circumstance of that era. Telecommunications companies were starting to develop optical switches for fiber-optic networks; they would use minuscule motors to change the angle of mirrors very quickly to route an optical signal in one direction or the other. Not long after I began reading Vogel and exploring elastic design, I embarked on a project with Steven Rodgers and his team at Sandia National Laboratories' microsystems division, where a monoform design seemed perfect.

Sandia needed to build a linear motor with sufficient output displacement to do work—at least 10 microns. Yet the fabrication constraints of electrostatic motors limit their motion to two microns. I knew I could not simply miniaturize, say, a geared transmission. Even if we could find someone with steady enough hands to assemble gears, hinges and shafts with dimensions in the one- to two-micron range, the resulting machine would be too sloppy for modern engineering. At MEMS scale, a machine with a tenth of a micron of clearance is about as useful as a Tinkertoy. Besides, MEMS devices are batch-fabricated much the same way as integrated circuits, tens of thousands in an area the size of a thumbnail. Given all that, I designed a monoform motion amplifier to generate 20 microns of output motion when integrated with the electrostatic motor.

Text C

Bottoms Up

Treated sewage could be the safest, most environmentally sound source of tap water yet—if we can get over the yuck factor.

By Olive Heffernan

San Diego imports as much as 90 percent of its water from the Colorado River to the east and the Sacramento–San Joaquin River Delta to the north. But both those sources are running dry. By converting effluent, San Diego could meet 40 percent of its daily water demands. And it would put an end to the city dumping poorly treated wastewater into the ocean.

But let's face it, not everyone wants a mouthful of treated sewage. This "yuck factor" quashed an attempt in the late 1990s to start a similar scheme in San Diego, and a poll in 2004 found that 63 percent of residents still opposed the idea of reuse. Numerous proposals in Australia have met the same fate, vetoed by vocal civic groups. Laurence Jones, who had founded one Australian group, Citizens Against Drinking Sewage, questions whether sewage sourced from hospitals, industry, homes and slaughterhouses can ever be fully cleaned.

Attitudes in San Diego have undergone an amazing turnaround, however, as drought has worsened and coastal neighborhoods have grown. Now nearly three quarters of the population are in favor of treated toilet water, but with one stipulation: That after the effluent is cleaned, it will be sent back to a reservoir, where it can be highly diluted and then treated further before being piped to homes.

That process is known as indirect potable reuse. The people running the Advanced Water Purification Facility, currently a test site for this approach, hope to take an additional step: treat effluent to a high level of purity and send it straight to the tap—known as direct potable reuse. For many residents, though, that last step goes too far. "It just seems more palatable to put the water back into a reservoir," says Megan Baehrens, executive director of San Diego Coast-keeper, a nonprofit organization that played a key role in persuading the city to launch the project.

Which process wins will determine what California regulators will allow San Diego, and the rest of the state, to do. And if direct reuse is sanctioned here, where environmental regulations are notoriously rigid, experts say the process will soon spread to other drought-afflicted communities worldwide.

● **The Cleaner the Better**

All eyes are indeed on the San Diego pilot facility. Right now the plant produces one million gallons of water a day. Although the water is purified to drinking standards, it is sent to irrigate the nearby Torrey Pines Golf Course and a cemetery. Steirer wants to scale up to 10 times the current capacity in the next five to 10 years. The default plan is to release the treated water into the local San Vicente Reservoir to dilute it, after which the mix of treated and reservoir water would be sanitized and sent to homes. Plan B—if regulators allow it—will be the direct approach.

Regulation will not be enough, however, for either approach to gain the public's nod. The utility must get consumers past the yuck factor. Critically, it must convince people that the water is clean. More than 4,000 visitors have toured the plant, among them mothers, Girl Scouts[1], doctors and elected officials. Many of them question the

---

[1] Girl Scouts refers to the Girl Scouts of the U.S.A. (GSUSA), a youth organization for girls in the U.S. and American girls living abroad. GSUSA aims to empower girls and to help teach values such as honesty, fairness, courage, compassion, character, sisterhood, confidence, entrepreneurship, and citizenship through activities including camping, community service, learning first aid, and earning badges by acquiring practical skills.

safety of consuming what was once raw sewage. It is not a trivial concern. Every year 19 million Americans become sick, and 900 of them die, from viruses, bacteria and parasites in water that has undergone the routine treatment that most municipalities use.

One way to win hearts and minds is to make sure the resulting water is purer than current water supplies are. On the tour, visitors learn that purified effluent is, ironically, much cleaner than their tap water is now. That is because most of us are drinking "downstream"—the river or lake that supplies our tap water doubles as a disposal site for water coming from standard sewage treatment plants, which is not clean enough to drink. "Water in the Mississippi River has been used five times by the time it reaches New Orleans," explains George Tchobanoglous, an international water expert at the University of California, Davis. Yet people expect water that will come from effluent to be held to a much higher standard than regular municipal supplies are.

Steirer says the purified wastewater in San Diego is indeed "much cleaner" than water that comes from a typical drinking water treatment plant. Furthermore, storing water for such a plant in a reservoir or an aquifer carries its own risks, says David Sedlak, engineering professor at U.C. Berkeley. Ducks and other animals introduce filth into reservoirs, for example, and arsenic can leach from rocks into groundwater.

Traditional treatment for U.S. drinking water goes through two or three steps for removing suspended solids and is then disinfected using chlorine. Transforming fragrant sewage into pristine tap water requires different engineering. The AWPF plant takes sewage water treated by the North City Water Reclamation Plant and adds higher levels of cleansing to "purify" it.

The first step at the AWPF is microfiltration, which happens in large tubes that resemble giant drums of pasta. Shane Trussell, president of Trussell Technologies and head of engineering at the project, tells me that each drum contains 9,000 of these pasta-like fibers and that each fiber is dotted with microscopic pores 300 times as narrow as a human hair. As water is forced through the tubes, the fibers filter out viruses, bacteria, protozoa and suspended solids.

Next, the water is sent at high pressure through tubes with even smaller fibers, in a process known as reverse osmosis❶. This step removes any remaining dissolved particles, up to 10,000 times as small as even the tiniest bacteria, including chemicals, viruses and pharmaceuticals. For the final step, water at the AWPF goes to advanced oxidation, where it is mixed in huge vats with minute amounts of highly concentrated hydrogen peroxide and then exposed to ultraviolet light. This stage destroys any remnant contaminants, even at quantities of parts per trillion, a dose equivalent a single drop in hundreds of Olympic swimming pools.

Of the one million gallons of wastewater entering the plant daily, 80 percent makes it through to final approval—it is as pure as premium-grade bottled brands. It could be sent to the San Vicente Reservoir if the plant had a permit for indirect reuse. As it stands, the water goes into the state's purple pipes, which supply the region with recycled water for irrigation and industry. The remaining 20 percent is sent to the local sewage treatment facility for disposal. Some of the substances that regularly turn up in the purified water are caffeine, hand cleaner and artificial sweetener, but they are in such minute doses as to be harmless, Trussell says. The final product is also extremely low in salt—20 parts per million (ppm), compared with 600 ppm in the city's imported water.

This past April, Trussell and his band of engineers added yet another step to make the water clearer and cleaner still. The purified water would be exposed to ozone, which would raise the removal of microbes, for example, to 99.9999 percent. The water would then go through specialized filters meant to further reduce any organic content. If successful, this single addition could be enough to convince regulators that there is no need to send the treated water to a reservoir. The water quality would far exceed all state and federal drinking-water standards; in fact, the purified water that was produced before this latest add-on step already met or exceeded those standards.

---

❶ Reverse osmosis is a technique for purifying water, in which pressure is applied to force liquid through a semipermeable membrane in the opposite direction to that in normal osmosis.

Unit 8
Engineering

● **Psychological Advantage**

Advocates of direct reuse need to overcome psychological resistance. Many people seem willing to consider indirect reuse partly because storing the water in a reservoir or aquifer provides an important psychological separation between sewage as the source and drinking water as the product. Lessons about acceptance can be learned from several communities that have successfully implemented indirect water reuse. In the late 1990s Orange County, California, faced diminishing water supplies. By 2008 it boasted the world's largest facility for supplementing local groundwater with treated effluent for drinking, processing 70 million gallons of wastewater daily, equivalent to 20 percent of local demand.

At the outset—much like in San Diego—Orange County residents were skeptical; 70 percent opposed the plan. But by the time the plant came online, it had the backing of the entire community, thanks to a very effective public relations campaign. Ron Wildermuth, who led the effort, explains that staff members of the county's water district had seven years' worth of data on water quality before they even approached the community. They then spent another 10 years talking to everyone from rotary and garden clubs to local businesses, explaining the options and inviting them to taste the water.

The effort set the stage for what is now happening in San Diego, which has adapted much of Orange County's technology, too. It hopes for the direct approach in part because it has no natural groundwater basins to store the purified water. Many municipalities across the U.S. and the world are in a similar situation, so San Diego is the proving ground.

Experiences in Australia reveal how not to influence the public. Some provinces have banned the drinking of reclaimed water altogether, and water reuse schemes for cities such as Brisbane and Melbourne—which are prone to long, sporadic droughts—have imploded under public opposition. The mistake made by the government was pushing for public acceptance at exactly the wrong time, Khan believes.

It is best to start the conversation early, Khan says, adding that now may be an opportune time to try again in Australia because water supplies have rebounded somewhat, providing some time for discussion with the public. One facility is ripe for conversion. Commissioned in 2006 at the height of a drought period, the Western Corridor Recycled Water Project is a $2.3-billion system that was originally developed to supply recycled water for industry, agriculture and drinking. The plan was to send the water to the Wivenhoe Dam, the source of most tap water in and around Brisbane. The system gathers effluent from six wastewater arguing that one of the advanced treatment plants should be converted to a direct reuse facility, which could fulfill about 35 percent of Brisbane's water supply.

If the Queensland government opts for this plan, it would create the largest direct potable reuse facility in the Southern Hemisphere. Convincing politicians and the public may be easier this time around, but they will need information in addition to time to consider the options—just like in Orange County.

Officials may be inspired by research published last year by the U.S.-based WateReuse Research Foundation. In a study, foundation researchers showed a group of Californians and Australians of mixed gender, age and education four scenarios for sourcing tap water. The first represented current practice and showed drinking water being sourced from a river that was also a disposal site for treated sewage. The second scenario showed cleansed sewage being further purified by a facility and then blended in a reservoir before being sent to a drinking-water plant for added treatment. In the third example, purified water was sent straight back to the river, where it mixed before being sent for treatment. In the final scenario—direct reuse—the purified water went straight to the city's homes, sidestepping the reservoir and the additional treatment plant. Study participants, irrespective of gender or education level, considered direct reuse as the safest option and the current practice as the least safe.

- **Necessity as Invention**

Demonstrating an utter lack of alternative water sources is another way to sway the public. That is what succeeded in Namibia, the only place in the world that is supplying directly recycled water on a significant scale. Back in 1957, severe drought depleted the city of Windhoek's groundwater supply in just eight weeks. Located about 190 miles inland and 500 miles from the closest perennial river, the community was left with no other reliable water source. By 1968 the city had a fully operational direct reuse facility. Today 25 percent of Windhoek's tap water comes from processed sewage. Windhoek faced fewer public challenges than San Diego does. For a start, "there was no activism back then," says Petrus du Pisani, who oversees the facility. "Citizens may have been a little wary, but they accepted the necessity of the decision." Back in the late 1960s, he notes "people had a lot of faith in science and officialdom". Still, the city informed locals and invited them to taste the water.

The system in Namibia would never fly elsewhere today, however. Although it uses multiple treatment steps, it does not include reverse osmosis, key to the San Diego project and others such as Orange County. Officials in Namibia say the water is safe and meets standards set by the World Health Organization. Windhoek, located inland, could not easily dispose of the large volume of waste brine that reverse osmosis creates. And in the 1960s "there were fewer man-made chemicals" in the wastewater, du Pisani says. "Our main concerns were soaps and frothing agents." One downside of omitting this step is that the drinking water is high in total dissolved solids, which makes it taste salty.

Du Pisani says Windhoek will probably add small-scale reverse osmosis by 2020 to reduce saltiness. Drinking-water standards are changing rapidly throughout the world, even in Namibia, he adds, noting that Windhoek's approach is no longer the most appropriate. The volume of brine, as well as the large amount of energy needed for reverse osmosis, could make direct potable reuse too costly for other communities. Ironically, new treatments are being developed that could reduce the amount of brine, and waste in general, from the overall process. Indirect reuse and desalination can both use reverse osmosis, too. Direct reuse is thus often less energy-intensive than those options because they require additional pipelines and energy to push water through them.

As drought has taken hold in the U.S., several communities have been forced to confront a fate similar to Windhoek's. Big Spring, Tex., had seen little rainfall, year after year. Cloudcroft, N.M., a small mountain community that more than doubles in size on weekends and holidays, had been hauling water over considerable distances.

In the past year both towns have begun to purify effluent to supplement their drinking water. Neither community boasts a suitable reservoir or aquifer for storing the treated water over the long term. Instead in Cloudcroft, the purified wastewater is blended with water taken from a local well or spring and stored temporarily in a holding tank before being treated again and piped to people's homes. In Big Spring, the cleansed wastewater is mixed with water from a distant, regional reservoir, and the mixture is treated.

● **A Taste for Success**

San Diego is not in such dire straits yet, which causes some experts to say the city should consider alternative solutions. Although "a big fan of the concept", water authority and Pacific Institute president Peter Gleick thinks that direct reuse is still decades away from happening in California. Gleick argues that California should instead focus on conserving water, both in cities and, even more so, in agricultural operations, which use 80 percent of the state's water supply.

Whether the AWPF plant is given a green light for direct or indirect reuse, it is a win for San Diego because it will be a reliable local supply of water, it will reduce waste dumped at sea and it will avoid the billions of dollars needed to upgrade the wastewater treatment plant. Until then, cleansed (but not purified) water will continue to run in the purple pipes that line the roads leading to industries. The pipes are marked clearly with signs that read, "Do not drink."

(Excerpt from *Scientific American*, July 2014)

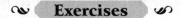

Exercises

📖 Reading Comprehension

Directions: Answer the following questions based on the information from the text.

1. How did San Diego residents respond to treated sewage in the late 1990's, 2004 and 2014 respectively? Tell how their attitude changed.

2. What are the indirect potable reuse of treated sewage and the direct reuse of treated sewage? Which approach is more popular with people? Why?
3. Describe the procedures of the sewage water treatment at the San Diego Advanced Water Purification Facility (AWPF) briefly.
4. Why is Namibia able to supply its capital of Windhoek with directly recycled water on a large scale? What is the downside of their recycled water?
5. According to the author, what is the significance of the AWPF?

Unit 9
Computer Science

导读

本单元的文章均涉及计算机科学技术及应用。Text A 详细讲解了一个充满传感器的世界将会如何改变我们所见、所闻、所思和所在的方式。Text B 详细讲述了新型计算机的开发历程。由于芯片技术的局限性，摩尔定律不再适合计算机芯片技术的发展，芯片制造商花费数十亿来开发基于新材料的处理器和计算机系统。Text C 详细阐述了比晶体管更接近神经元的新型电子元器件如何形成效率更高和速度更快的"忆计算"。

Part I Intensive Reading

Text A

Extra Sensory Perception

How a world filled with sensors will change the way we see, hear, think and live.

By Gershon Dublon and Joseph A. Paradiso

Here's a fun experiment: Try counting the electronic sensors surrounding you right now. There are cameras and microphones in your computer; GPS sensors and gyroscopes in your smartphone; Accelerometers in your fitness tracker. If you work in a modern office building or live in a newly renovated house, you are constantly in the presence of sensors that measure motion, temperature and humidity.

Sensors have become abundant because they have, for the most part, followed Moore's law❶: They just keep getting smaller, cheaper and more powerful. Thanks

---

❶ Moore's law is the observation that the number of transistors in a dense integrated circuit doubles approximately every two years. The observation is named after Gordon Moore, the co-founder of Fairchild Semiconductor and Intel.

to progress in microelectronics design as well as management of energy and the electromagnetic spectrum, a microchip that costs less than a dollar can now link an array of sensors to a low-power wireless communications network.

The amount of information this vast network of sensors generates is staggering—almost incomprehensible. Yet most of these data are invisible to us. Today sensor data tend to be "siloed", accessible by only one device for use in one specific application, such as controlling your thermostat or tracking the number of steps you take in a day.

Eliminate these silos, and computing and communications will change in profound ways. Once we have protocols that enable devices and applications to exchange data, sensors in anything can be made available to any application. When that happens, we will enter the long-predicted era of ubiquitous computing.

Predicting what ubiquitous computing and sensor data will mean for daily life is as difficult as predicting 30 years ago how the Internet would change the world. Fortunately, media theory can serve as a guide. In the 1960s communications theorist Marshall McLuhan spoke of electronic media, mainly television, becoming an extension of the human nervous system. When sensors are everywhere—and when the information they gather can be grafted onto human perception in new ways—where do our senses stop? What will "presence" mean when we can funnel our perception freely across time, space and scale?

- **Visualizing Sensor Data**

We perceive the world using all our senses, but we digest most digital data through tiny two-dimensional screens on mobile devices. It is no surprise, then, that we are stuck in an information bottleneck. As the amount of information about the world explodes, we find ourselves less able to remain present in that world. Yet there is a silver lining❶ to this abundance of data, as long as we can learn to use it properly. That is why our group at the M.I.T. Media Lab has been working for years on ways

---

❶ A silver lining is the latter part of the proverb—every cloud has a silver lining, meaning that every difficult or sad situation has a comforting or more hopeful aspect, even though this may not be immediately apparent.

to translate information gathered by networks of sensors into the language of human perception.

Just as browsers like Netscape gave us access to the mass of data contained on the Internet, so will software browsers enable us to make sense of the flood of sensor data that is on the way. So far the best tool for developing such a browser is the video game engine—the same software that lets millions of players interact with one another in vivid, ever changing three-dimensional environments. Working with the game engine Unity 3D, we have developed an application called DoppelLab that takes streams of data collected by sensors placed throughout an environment and renders the information in graphic form, overlaying it on an architectural computer-aided design (CAD) model of the building. At the Media Lab, for example, DoppelLab collects data from sensors throughout the building and displays the results on a computer screen in real time. A user looking at the screen can see the temperature in every room, or the foot traffic in any given area, or even the location of the ball on our smart Ping-Pong table.

DoppelLab can do much more than visualize data. It also gathers sounds collected by microphones scattered about the building and uses them to create a virtual sonic environment. To guarantee privacy, audio streams are obfuscated at the originating sensor device, before they are transmitted. This renders speech unintelligible while maintaining the ambience of the space and the vocal character of its occupants. DoppelLab also makes it possible to experience data recorded in the past.

Sensor browsers such as DoppelLab have immediate commercial applications—for example, as virtual-control panels for large, sensor-equipped buildings. In the past a building manager who wanted to track down a problem in the heating system might have sorted through spreadsheets and graphs, cataloguing anomalous temperature measurements and searching for patterns that would point to the source. Using DoppelLab, that person can see the current and desired temperature in every room at once and quickly spot issues that span multiple rooms or floors. More than that, planners, designers and building occupants alike can see how the infrastructure is being used.

Unit 9
Computer Science

But we did not make DoppelLab with commercial potential in mind. We built it to explore a bigger and more intriguing matter: the impact of ubiquitous computing on the basic meaning of presence.

• Redefining Presence

When sensors and computers make it possible to virtually travel to distant environments and "be" there in real time, "here" and "now" may begin to take on new meanings. We plan to explore this shifting concept of presence with DoppelLab and with a project called the Living Observatory at Tidmarsh Farms, which aims to immerse both physical and virtual visitors in a changing natural environment.

Since 2010 a combination of public and private environmental organizations have been transforming 250 acres of cranberry bogs in southern Massachusetts into a protected coastal wetland system. The bogs, collectively called Tidmarsh Farms, are co-owned by one of our colleagues, Glorianna Davenport. Having built her career at the Media Lab on the future of documentary, Davenport is fascinated by the idea of a sensor-rich environment producing its own "documentary". With her help, we are developing sensor networks that document ecological processes and enable people to experience the data those sensors produce. We have begun populating Tidmarsh with hundreds of wireless sensors that measure temperature, humidity, moisture, light, motion, wind, sound, tree sap flow and, in some cases, levels of various chemicals.

Efficient power management schemes will enable these sensors to live off their batteries for years. Some of the sensors will be equipped with solar cells, which will provide enough of a power boost to enable them to stream audio—the sound of the breeze, of nearby birds chirping, of raindrops falling on the surrounding leaves. Our geosciences colleagues at the University of Massachusetts Amherst are outfitting Tidmarsh with sophisticated ecological sensors, including submersible fiber-optic temperature gauges and instruments that measure dissolved oxygen levels in the water. All these data will flow to a database on our servers, which users can query and explore with a variety of applications.

Some of these applications will help ecologists view environmental data collected at the marsh. Others will be designed for the general public. For example, we are developing a DoppelLab-like browser that can be used to virtually visit Tidmarsh from any computer with an Internet connection. In this case, the backdrop is a digital rendering of the topography of the bog, filled with virtual trees and vegetation. The game engine adds noises and data collected by the sensors in the marsh. Sound from the microphone array is blended and cross-faded❶ according to a user's virtual position; you will be able to soar above the bog and hear everything happening at once, listen closely to a small region, or swim underwater and hear sound collected by hydrophones. Virtual wind driven by real-time data collected from the site will blow through the digital trees.

The Living Observatory is more of a demonstration project than a practical prototype, but real-world applications are easy to imagine. Farmers could use a similar system to monitor sensor-laden plots, tracking the flow of moisture, pesticides, fertilizers or animals in and around their cropland. City agencies could use it to monitor the progression of storms and floods across a city while finding people in danger and getting them help. Eventually this kind of remote presence could provide the next best thing to teleportation: Travelers could project themselves into their homes to spend time with their families while on the road.

● Augmenting Our Senses

It is a safe bet that wearable devices will dominate the next wave of computing. We view this as an opportunity to create much more natural ways to interact with sensor data. Wearable computers could, in effect, become sensory prostheses.

Researchers have long experimented with wearable sensors and actuators on the body as assistive devices, mapping electrical signals from sensors to a person's existing senses in a process known as sensory substitution. Recent work suggests that neuroplasticity—the ability of our brain to physically adapt to new stimuli—may enable perceptual-level cognition of "extra sensory" stimuli delivered through our existing sensory channels. Yet there is still a huge gap between sensor network data and human sensory experience.

---

❶ Cross-faded means to make a picture or sound appear or be heard gradually as another disappears or becomes silent.

We believe one key to unlocking the potential of sensory prostheses will be gaining a better handle on the wearer's state of attention. Today's highest-tech wearables, such as Google Glass, tend to act as third-party agents on our shoulders, suggesting contextually relevant information to their wearer (recommending a particular movie as a wearer passes a movie theater, for example). But these suggestions come out of the blue. They are often disruptive, even annoying, in a way that our sensory systems would never be. Our sensory systems allow us to tune in and out dynamically, attending to stimuli if they demand it but otherwise focusing on the task at hand. We are conducting experiments to see if wearable computers can tap into the brain's inherent ability to focus on tasks while maintaining a preattentive connection to the environment.

Our first experiment will determine whether a wearable device in the field can pick out which of a set of audio sources a user is listening to. We would like to use this information to enable the wearer of a device to tune into the live microphones and hydrophones at Tidmarsh in much the same way that they would tune into different natural sources of sounds. This approach to delivering digital information could mark the beginning of a fluid connection between our sensory systems and networked sensor data. We hope these devices, and the information they provide, will fold into our existing systems of sensory processing rather than further displacing them.

- **Dream or Nightmare?**

For many people, ourselves included, the world we have just described has the potential to be frightening. Redefining presence means changing our relationship with our surroundings and with one another. Even more concerning, ubiquitous computing has tremendous privacy implications. Yet we believe there are many ways to build safeguards into technology.

One approach is to make sensors respond to context and a person's preferences. Nan-Wei Gong explored an idea of this kind when she was with our research group several years ago. She built a special key fob that emitted a wireless beacon informing nearby sensor devices of its user's personal privacy preferences. Each badge had a large button labeled "No"; on pressing the button, a user was guaranteed an interval of total privacy wherein all sensors in range were blocked from transmitting his or her data.

Any solution will have to guarantee that all the sensor nodes around a person both receive and honor such requests. Designing such a protocol presents technical and legal challenges. Yet research groups around the world are already studying various approaches to this conundrum. For example, the law could give a person ownership or control of data generated in his or her vicinity; a person could then choose to encrypt or restrict those data from entering the network. One goal of both DoppelLab and the Living Observatory is to see how these privacy implications play out in the safe space of an open research laboratory.

Meanwhile we will be able to start seeing what kinds of new experiences await us in a sensor-driven world. We think it is entirely possible to develop technologies that will fold into our surroundings and our bodies. These tools will get our noses off the smartphone screen and back into our environments. They will make us more, rather than less, present in the world around us.

(Excerpt from Scientific American, July 2014)

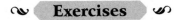

I. Reading Comprehension

● Section One

Directions: Answer the following questions based on the information from the text.

1. What does the era of ubiquitous computing mean?
2. What can people do with the application of DopperLab?
3. What did the researchers do to explore the shifting concept of presence?
4. What roles will wearable computers and sensor data play in our daily life, and in what way are the roles going to be played out?
5. What problems does ubiquitous computing have? Introduce some possible solutions to these problems.

Unit 9
Computer Science

● **Section Two**

Directions: Write an abstract based on the text in no more than 200 words.

Abstract:
Key words:

II. Vocabulary

● **Section One**

Directions: Choose the word or phrase that is closest in meaning to the underlined part in each sentence.

1. When sensors are everywhere—and when the information they gather can be <u>grafted</u> onto human perception in new ways—where do our senses stop?

    A. inserted or fixed
    B. worked
    C. made money by dishonest means
    D. transplanted a tissue

2. What will "presence" mean when we can <u>funnel</u> our perception freely across time, space and scale?

    A. direct something through a narrow space
    B. pour something into a narrow space
    C. assume a conical shape
    D. send information from one place to another

3. We perceive the world using all our senses, but we <u>digest</u> most digital data through tiny two-dimensional screens on mobile devices.

    A. break down food
    B. understand or assimilate
    C. arrange and integrate in the mind
    D. put up with something unpleasant

4. That is why our group at the M.I.T. Media Lab has been working for years on ways to <u>translate</u> information gathered by networks of sensors into the language of human perception.

   A. restate (words) from one language into another language

   B. change from one form into another

   C. change the position of (figures or bodies) in space without rotation

   D. move from one place to another

5. Just as browsers like Netscape gave us access to the mass of data contained on the Internet, so will software browsers enable us to make sense of the <u>flood</u> of sensor data that is on the way.

   A. the rising of a body of water and its overflowing onto normally dry land

   B. an outpouring of tears or emotion

   C. a very large quantity

   D. the inflow of the tide

6. We plan to explore this shifting concept of presence with DoppelLab and with a project called the Living Observatory at Tidmarsh Farms, which aims to immerse both <u>physical</u> and virtual visitors in a changing natural environment.

   A. of or relating to the body as opposed to the mind

   B. relating to the sciences dealing with matter and energy, especially physics

   C. of things perceived through the senses as opposed to the mind, tangible or concrete

   D. relating to the structure, size, or shape of something that can be touched and seen

7. We have begun <u>populating</u> Tidmarsh with hundreds of wireless sensors that measure temperature, humidity, moisture, light, motion, wind, sound, tree sap flow and, in some cases, levels of various chemicals.

   A. forming the population of (a place)   B. causing people to settle in

   C. filling or being present in   D. inhabiting

8. Researchers have long experimented with wearable sensors and actuators on the body as assistive devices, <u>mapping</u> electrical signals from sensors to a person's existing senses in a process known as sensory substitution.

   A. recording in detail the spatial distribution of

   B. representing an area of land or sea showing physical features

C. associating each element of one set with an element of another set

D. planning or arranging something in detail

9. Even more concerning, ubiquitous computing has tremendous privacy <u>implications</u>.

   A. conclusions that can be drawn from something although it is not explicitly stated

   B. likely consequences of something

   C. the actions of being involved in something

   D. things implied

10. Any solution will have to guarantee that all the sensor nodes around a person both receive and <u>honor</u> such requests.

    A. regard with great respect   B. grace or privilege

    C. fulfill or keep   D. accept or pay when due

● **Section Two**

Directions: There are two or three meanings for each semi-technical term underlined in the following sentences. Choose the correct one according to the context.

1. There are cameras and microphones in your computer. GPS sensors and gyroscopes in your smartphone. Accelerometers in your fitness <u>tracker</u> (a person who tracks someone or something by following their trails; a device that follows and records the movements of someone or something; a connecting rod in the mechanism of an organ).

2. Eliminate these <u>silos</u> (tall towers or pits on a farm used to store grain; underground chambers in which a guided missile is kept ready for firing; systems, processes, departments etc. that operate in isolation from others), and computing and communications will change in profound ways.

3. Once we have protocols that enable devices and <u>applications</u> (formal requests to an authority for something; actions of putting something into operation; programs of software designed and written to fulfill particular purposes of the user) to exchange data, sensors in anything can be made available to any application.

4. Some of the sensors will be equipped with solar cells, which will provide enough of a power boost to enable them to <u>stream</u> (run or flow in a continuous current in a specified direction; move in a continuous flow in a specified direction; relay over the Internet as a steady, continuous flow) audio.

5. Sound from the microphone array is blended and cross-faded according to a user's virtual (almost or nearly as described, but not completely or according to strict definition; not physically existing as such but made by software to appear to do so) position.

6. Farmers could use a similar system (a set of connected things or parts forming a complex whole, in particular; a set of principles or procedures according to which something is done; a group of related hardware units or programs or both, especially when dedicated to a single application) to monitor sensor-laden plots, tracking the flow of moisture, pesticides, fertilizers or animals in and around their cropland.

7. With her help, we are developing sensor networks that document (record something in written, photographic, or other forms; support or accompany with documentation) ecological processes and enable people to experience the data those sensors produce.

8. Any solution will have to guarantee that all the sensor nodes (points in a network or diagram at which lines or pathways intersect or branch; the parts of a plant stem from which one or more leaves emerge, often forming a slight swelling or knob) around a person both receive and honor such requests.

9. Working with the game engine Unity 3D, we have developed an application called DoppelLab that takes streams of data and renders the information in graphic (of or relating to visual art, especially involving drawing, engraving, or lettering; of or relating to, or denoting a visual image) form, overlaying it on an architectural computer-aided design model of the building.

10. DoppelLab can do much more than visualize (form a mental image of; make something visible to the eye) data.

● **Section Three**

Directions: Match the Chinese terms with their English equivalents.

1. 普适计算　　　　　　　　A. accelerometer
2. 游戏引擎　　　　　　　　B. hydrophone
3. 计算机辅助设计　　　　　C. perception language
4. 媒介理论　　　　　　　　D. presence
5. 通信理论学家　　　　　　E. communications theorist

Unit 9
Computer Science

6. 存在感   F. game engine
7. 感官语言   G. computer-aided design
8. 虚拟控制平台   H. wearable device
9. 水下光纤温度计   I. neuroplasticity
10. 水下测音器   J. submersible fiber-optic temperature gauge
11. 加速计   K. gyroscope
12. 可穿戴设备   L. media theory
13. 感知替换   M. ubiquitous computing
14. 神经可塑性   N. virtual-control panel
15. 陀螺仪   O. sensory substitution

III. Questions for Discussion

Directions: Work in groups and discuss the following questions.

1. What roles do sensors play in our daily life?
2. Do you want to be "seen" in a world full of sensors? Why or why not?
3. Will computer take the place of human brain? Support your opinion with sound facts or relevant examples.

Part II Extensive Reading

Text B

The Search for a New Machine

With the end of Moore's law in sight, chip manufacturers are spending billions to develop novel computing technologies.

By John Pavlus

In a tiny, windowless conference room at the R&D headquarters of Intel, Mark Bohr, the company's director of process architecture and integration, is coolly explaining how Moore's law is dead—and has been for some time. Bohr says: "You have to understand that the era of traditional transistor scaling, where you take the same basic structure and materials and make it smaller—ended about 10 years ago."

In 1965 Gordon Moore, then director of R&D at Fairchild Semiconductor, published the bluntly entitled document "Cramming More Components onto Integrated Circuits". Moore predicted that the number of transistors that could be built into a chip at optimal cost would double every year. A decade later he revised his prediction into what became known as Moore's law: Every two years the number of transistors on a computer chip will double.

Integrated circuits make computers work. But Moore's law makes computers evolve. Because transistors are the "atoms" of electronic computation—the tiny switches that encode every 1 and 0 of a computer's memory and logic as a difference in voltage—if you double the number of them that can fit into the same amount of physical space, you can double the amount of computing you can do for the same cost. Intel's first general-purpose microprocessor, the 8080, helped to launch the PC revolution when it was released in 1974. The two-inch-long, candy bar–shaped wafer contained 4,500 transistors. As of this writing, Intel's high-performance server central processing units (CPUs) contain 4.5 billion transistors each.

But even Moore's law buckles under the laws of physics—and within a decade it will no longer be possible to maintain this unprecedented pace of miniaturization. That is why chip manufacturers such as Intel, IBM and Hewlett-Packard (HP) are dumping billions of R&D dollars into figuring out a post–Moore's law world.

HP is going even further; it wants to extend the fundamental theory of electronics itself. The company has built a prototype computer, code-named "the Machine", that leverages the power of a long-sought missing link of electronics: the memristor. This component allows the storage and random-access memory (RAM) functions of computers to be combined. The common metaphor of the CPU as a computer's "brain" would become more accurate with memristors instead of transistors because the former actually work more like neurons: They transmit and encode information as well as store it. Combining volatile memory and nonvolatile storage in this way could dramatically increase efficiency and diminish the so-called von Neumann bottleneck, which has constrained computing for half a century.

None of these technologies are ready to replace, or even augment, the chips in our laptops or phones in the next few years. But by the end of the decade at least one of them must be able to deliver computational performance gains that have a chance of taking over where traditional silicon circuit engineering inevitably trails off. The question is: Which one—and when?

● **Beyond Silicon**

The idea behind Moore's law is simple—halving the size of a transistor means you can get double the computing performance for the same cost. Gordon Moore's 1965 paper may have predicted what would happen to transistor density every other year, but he never described how performance doubling would emerge from that increased density. It took another nine years for a scientist at IBM named Robert Dennard to publish an explanation now known as Dennard scaling. It describes how the power density of MOSFETs[1] stays constant as their physical size scales down. In other words, as transistors shrink, the amount of electric voltage and current required to switch them on and off shrinks, too.

---

[1] MOSFET stands for metal-oxide-semiconductor field-effect transistor.

For 30 years Dennard scaling was the secret driver of Moore's law, guaranteeing the steady PC performance gains that helped people start businesses, design products, cure diseases, guide spacecraft and democratize the Internet. And then it stopped working. Once fabs began etching features into silicon smaller than 65 nanometers (about half the length of an HIV virus), chip designers found that their transistors began to "leak" electrons because of quantum-mechanical effects. The devices were simply becoming too tiny to reliably switch between "on" and "off". Not only that, researchers at IBM and Intel were discovering a so-called frequency wall that put a limit on how fast silicon-based CPUs could execute logical operations—about four billion times per second—without melting down from excess heat.

Since 2000, chip engineers faced with these obstacles have been developing clever work-arounds. They have dodged the frequency wall by introducing multicore CPUs (a 10-gigahertz processor will burn itself up, but four, eight or 16 3-GHz processors working together will not). They have shored up leaky transistors with "tri-gates" that control the flow of current from three sides instead of one. And they have built systems that let CPUs outsource particularly strenuous tasks to special-purpose sidekicks. But these stopgaps will not change the fact that silicon scaling has less than a decade left to live.

That is why some chipmakers are looking for ways to ditch silicon. Last year IBM announced that it was allocating $3 billion to aggressively research various forms of postsilicon computing. The primary material under investigation is graphene: sheets of carbon just one atom thick. Like silicon, graphene has electronically useful properties that remain stable under a wide range of temperatures. Even better, electrons zoom through it at relativistic speeds. And most crucially, it scales—at least in the laboratory. Graphene transistors have been built that can operate hundreds or even thousands of times faster than the top-performing silicon devices, at reasonable power density, even below the five-nanometer threshold in which silicon goes quantum.

Yet unlike silicon, graphene lacks a "bandgap": The energy difference between orbitals in which electrons are bound to the atom and those in which the electrons are free to move around and participate in conduction. Without a bandgap, it is very difficult to stem the flow of current that turns a transistor from on to off—which means that a graphene device cannot reliably encode digital logic. Carbon nanotubes may hold more promise. When sheets of graphene are rolled into hollow cylinders, they can acquire a small bandgap that gives them semiconducting properties akin to what silicon has, reopening the possibility of using them for digital transistors.

But carbon nanotubes are delicate structures. If a nanotube's diameter or chirality—the angle at which its carbon atoms are "rolled"—varies by even a tiny amount, its bandgap may vanish, rendering it useless as a digital circuit element. Engineers must also be able to place nanotubes by the billions into neat rows just a few nanometers apart, using the same technology that silicon fabs rely on now.

● **Breaking Down the Memory Wall**

"What's the most expensive real estate on the planet?" Andrew Wheeler asks. "This, right here." He points to a box drawn in black marker on a whiteboard, representing the die of a microchip. As the deputy director of HP Labs, Wheeler is explaining what most of the transistors occupying that premium real estate are actually used for. It is not for computation, he says. It is called "cache memory" or static RAM (SRAM), and all it does is store frequently accessed instructions. It is the silicon equivalent of the dock on your Mac—the place where you keep things you want to avoid digging for. Wheeler wants it to disappear. But he is getting ahead of himself. In the near term, he will settle for getting rid of your computer's hard drive and main memory.

According to HP, these three items—collectively known as the memory hierarchy, with SRAM at the top and hard drives at the bottom—are responsible for most of the problems faced by engineers grappling with Moore's law. Without high-speed, high-capacity memory to store bits and ship them as quickly as possible, faster CPUs do little good.

To break down this "memory wall", Wheeler's team in Palo Alto, Calif., has been designing a new kind of computer—the Machine—that avoids the memory hierarchy altogether by collapsing it into one unified tier. Each tier in the memory hierarchy is good at some things and bad at others. SRAM is extremely fast but power-hungry and low-capacity. Main memory, or dynamic RAM (DRAM), is pretty fast, dense and durable—which is good, because this is the workbench that your computer uses to run active applications. Of course, cutting the power makes everything in DRAM disappear, which is why "nonvolatile" storage media such as flash and hard disks are necessary for saving data in long-term storage. They offer high capacity and low power consumption, but they are glacially slow (and flash memory wears out relatively quickly). Because each memory medium has overlapping trade-offs, modern computers link them together so that CPUs can shuttle data up and down the tiers as efficiently as possible.

A universal memory that could combine the speed of SRAM, the durability of DRAM, and the capacity and power efficiency of flash storage has been a holy grail for engineers, designers and programmers for decades, Wheeler says. The Machine exploits an electronic component, the memristor, to cover the latter two items on the universal-memory wish list. Mathematically predicted in 1971, the memristor—which is a blend of the words "memory resistor" because the device's ability to conduct electricity depends on the amount of current that previously flowed through it—was long believed to be impossible to build. In 2008 HP announced that it had constructed a working memristor; the research program was internally fast-tracked and became the precursor of the Machine.

Pulsing a memristor memory cell with voltage can change its conductive state, creating the clear on/off distinction necessary for storing digital data. As with flash memory, that state persists when the current is removed. And like DRAM, the cells can be read and written at high speed while densely packed together. To achieve SRAM-like performance, though, memristor cells would need to be placed adjacent to the CPU on the same die of silicon to connect its high-performance memristor memory to the standard SRAM caches on logic processors. It is not quite the holy grail of universal memory, but it is close.

By combining RAM with nonvolatile storage, a memristor-based architecture like the Machine could enable massive increases in computer performance without relying on Moore's law. The version of IBM's Watson supercomputer that beat human contestants on *Jeopardy* in 2011 needed 16 terabytes of DRAM—housed in

10 power-guzzling Linux server racks—to perform its feats in real time. The same amount of nonvolatile flash memory could fit into a shoe box while consuming the same amount of power as an average laptop. A memory architecture that combined both functions at once would allow enormous data sets to be held in active memory for real-time processing rather than diced into smaller, sequential chunks—and at much lower energy costs.

As more and more connected devices join the "Internet of Things", the problem of streaming countless petabytes of information to and from data centers for storage and processing will make Moore's law moot, Wheeler says. Yet if universal memory enables supercomputerlike capabilities in smaller, less energy-hungry packages, these data streams could be stored and preprocessed locally by the connected devices themselves. Silicon CPU elements might never get smaller than seven nanometers or faster than four gigahertz—but with the memory wall torn down, it may no longer matter.

• Beyond von Neumann

Even if HP succeeds in its gambit to build universal memory, computers will still remain what they have always been since ENIAC, the first general-purpose computer, was built in 1946: extremely fast numerical calculators. Their essential design, formalized by mathematician John von Neumann in 1945, consists of a processing unit to execute instructions, a memory bank to store those instructions and the data they are to operate on, and a connection, or "bus", linking them. This von Neumann architecture is optimized for executing symbolic instructions in a linear sequence—also known as doing arithmetic.

But today we increasingly need computers to do jobs that do not map well to linear mathematical instructions: tasks such as recognizing objects of interest in hours of video footage or guiding autonomous robots through unstable or dangerous territory. Organisms must extract actionable information from a dynamic environment in real time; if a housefly were forced to pass discrete instructions back and forth, one by one, between separate memory and processor modules in its brain, it would never complete the computation in time to avoid getting swatted.

Dharmendra Modha, founder of IBM's cognitive computing group, wants to build computer chips that are at least as "smart" as that housefly—and as energy-efficient. The key, he explains, has been to scrap the calculatorlike von Neumann architecture. Instead his team has aimed to mimic cortical columns in the mammalian brain, which process, transmit and store information in the same structure, with no bus bottlenecking the connection. IBM's recently unveiled TrueNorth chip contains more than five billion transistors arranged into 4,096 neurosynaptic cores that model one million neurons and 256 million synaptic connections.

What that arrangement buys is real-time pattern-matching performance on the energy budget of a laser pointer. Modha points to a video monitor in the corner of the demo room at the IBM Almaden research campus in San Jose, Calif. The scene on it looks like surveillance footage from a camera that needs a hard reboot: Cars, pedestrians, and a bicycle or two are frozen in place on a traffic roundabout; one of the pedestrians is highlighted by a red box superimposed on the image. After a minute, the cars, people and bikes lurch into a different frozen position, as if the footage suddenly skipped ahead.

"You see, it's not a still image," Modha explains. "That's a video stream from Stanford's campus being analyzed by a laptop simulating a TrueNorth chip. It's just running about 1,000 times slower than real time." The actual TrueNorth chip that usually runs the video stream was being used for an internal training session in an auditorium next door, so I was not witnessing the chip's real performance. If I were, Modha says, the video feed would be playing at real-time speed, and the little red boxes would smoothly track the pedestrians as they entered and exited the frame.

Just like their von Neumann architecture counterparts, neurosynaptic devices such as TrueNorth have their own inherent weaknesses. "You wouldn't want to run iOS with this chip," Modha says. "I mean, you could, but it would be horribly inefficient—just like the laptop is inefficient at processing that video stream." IBM's goal is to harness the efficiencies of both architectures—one for precise and logical calculation, the other for responsive, associative pattern matching—into what it describes as a holistic computing system.

In this vision, the classic formulation of Moore's law still matters. Modha's team has already tiled 16 TrueNorth chips into a board, and by the end of this year the group plans to stack eight boards together into a 100-watt, toaster oven–sized device whose computational power "would require an entire data center to simulate".

In other words, silicon and transistor counts still matter—but what matters more is how they are arranged. Modha says. "A lot of people believed that you really needed to change the technology to get the gains. In fact, it became clear that while new technology may bring gains, a new architecture gave orders-of-magnitude gains in performance that were easy pickings in comparison."

- **Moore's Laws**

Back at building RA3 in Hillsboro, Michael C. Mayberry, Intel's director of components research, is dispelling another myth about Moore's law: It was never really about transistors. "Cost per function is the game," he says. Whether it is measured in transistors per square centimeter of silicon, instructions of code executed per second, or performance per watt of power, all that matters is doing ever more work with ever fewer resources. It is no surprise that on its own Web site, Intel describes Moore's law not as a technological trend or force of nature but as a business model.

"When Dennard scaling ended, that doesn't mean we stopped. We just changed. If you look forward 15 years, we can see several changes coming, but it doesn't mean we're going to stop." Mayberry says. What Intel, IBM and HP all agree on is that the future of computational performance—that is, how the industry will collectively deliver increased function at decreased cost—will cease to look like a line or a curve and will instead look more like the multibranching tree of biological evolution itself.

That is because our vision of computers themselves is evolving. What is really dying is not Moore's law but the era of efficient general-purpose computation that Moore's law described and enabled. Instead the relentless pursuit of lower cost per function will be driven by so-called heterogeneous computing, as Moore's law splits into Moore's laws. Companies such as IBM, Intel, HP and others will integrate not just circuits but entire systems that can handle the multiplying demands of distinct computational workloads. Bernard S. Meyerson of IBM says that people buy functions, not computer chips; indeed, they are less and less interested in buying computers at all.

Futurists such as Nick Bostrom (author of *Superintelligence: Paths, Dangers, Strategies*) presume that Moore's law will cause generalized artificial intelligence to take off and coalesce into a kind of all-knowing, omnipotent digital being. But heterogeneous computing suggests that computation is more likely to diffuse outward into formerly "dumb" objects, systems and niches—imbuing things such as cars, network routers, medical diagnostic equipment and retail supply chains with the semiautonomous flexibility and context-specific competence of domestic animals. In other words, in a post–Moore's law world, computers will not become gods—but they will act like very smart dogs.

And just as a Great Dane is not built to do the job of a terrier, a graphics processor is not built to do the work of a CPU. HP's Wheeler offers the example of multiple special-purpose processing cores "bolted onto" a petabyte-scale pool of universal memory—a hybrid of processing power and massive memory that works in much the same way that dedicated graphics accelerators and memory caches are marshaled around centralized CPU resources now. Meanwhile IBM's Modha envisions golf ball-size devices, consisting of cognitive chips fastened to cheap cameras, which could be dropped into natural disaster sites to detect highly specific patterns such as the presence of injured children. Computer scientist Leon Chua of the University of California, Berkeley, who first theorized the existence of memristors in 1971, says that HP's efforts to collapse the memory hierarchy and IBM's research on reimagining the CPU are complementary responses to what he calls "the Great Data Bottleneck". "It's incredible that the computers we've been using for everything for the past 40 years are all still based on the same idea" of the calculator-like von Neumann architecture, he says. The two-front transition to heterogeneous computing is "inevitable", he asserts, and "will create an entirely new economy"—not least because post-Moore's law, post-von Neumann computing will require entirely new methods of programming and designing systems. So much of modern computer science, engineering and chip design is concerned with masking the inherent limitations that the memory hierarchy and von Neumann architecture impose on computation,

Unit 9
Computer Science

Chua says that, once those limitations are removed, "every computer programmer will have to go back to school."

What Chua, Modha and Wheeler never mention in these near-future visions are transistors—or the predictable performance gains that the world has come to expect from them. According to IBM's Meyerson, what Moore's law has accurately described for half a century—a tidy relation between increased transistor density and decreased cost per function—may turn out to be a temporary coincidence. "If you look at the past 40 years of semiconductors, you can see a very constant heartbeat," Meyerson says. "It's not that progress won't continue. But this technology has now developed an arrhythmia."

(Excerpt from *Scientific American*, May 2015)

Exercises

I. Translate the following technical terms into English.

1. 集成电路 _____
2. 中央处理器 _____
3. 忆阻器 _____
4. 晶体管 _____
5. 三闸晶体管 _____
6. 计算性能 _____
7. 缩放定律 _____
8. 外包 _____
9. 高速缓冲存储器 _____
10. 能耗大 _____
11. 随机存取存储器 _____
12. 量子力学效应 _____
13. 逻辑运算 _____
14. 录像片段 _____
15. 闪速存储器 _____
16. 内存条 _____
17. 异构计算 _____
18. 分级存储器体系 _____
19. 神经突触核心 _____
20. 线性序列 _____

II. Translate the following paragraphs into Chinese.

According to HP, these three items—collectively known as the memory hierarchy, with SRAM at the top and hard drives at the bottom—are responsible for most of the problems faced by engineers grappling with Moore's law. Without high-speed, high-capacity memory to store bits and ship them as quickly as possible, faster CPUs do little good.

To break down this "memory wall", Wheeler's team in Palo Alto, Calif., has been designing a new kind of computer—the Machine—that avoids the memory hierarchy altogether by collapsing it into one unified tier. Each tier in the memory hierarchy is good at some things and bad at others. SRAM is extremely fast (so it can keep up with the CPU) but power-hungry and low-capacity. Main memory, or dynamic RAM (DRAM), is pretty fast, dense and durable—which is good, because this is the workbench that your computer uses to run active applications. Of course, cutting the power makes everything in DRAM disappear, which is why "nonvolatile" storage media such as flash and hard disks are necessary for saving data in long-term storage. They offer high capacity and low power consumption, but they are glacially slow (and flash memory wears out relatively quickly). Because each memory medium has overlapping trade-offs, modern computers link them together so that CPUs can shuttle data up and down the tiers as efficiently as possible. "It's an absolute marvel of engineering," Wheeler says. "But it's also a huge waste."

Text C

Just Add Memory

New types of electronic components, closer to neurons than to transistors, are leading to tremendously efficient and faster "memcomputing".

By Massimiliano Di Ventra and Yuriy V. Pershin

When we wrote the words you are now reading, we were typing on the best computers that technology now offers: machines that are terribly wasteful of energy and slow when tackling important scientific calculations. And they are typical of every computer that exists today, from the smartphone in your hand to the multimillion-dollar supercomputers humming along in the world's most advanced

Unit 9
Computer Science

computing facilities.

We were writing in Word, a perfectly fine program that you probably use as well. To write "When we wrote the words you are now reading", our computer had to move a collection of 0's and 1's from a temporary memory area and send it to another physical location, the CPU, via a bunch of wires. The processing unit transformed the data into the letters that we saw on the screen. To keep that particular sentence from vanishing once we turned our computer off, the data representing it had to travel back along that bunch of wires to a more stable memory area such as a hard drive.

This two-step shuffle happens because computer memory cannot do processing, and processors cannot store memory. It is a standard division of labor, and it happens even in fancy computers that do the fastest kind of calculating, called parallel processing, with multiple processors. The trouble is that each of these processors is still hobbled by this limitation. Scientists have been developing a way to combine the previously uncombinable: to create circuits that juggle numbers and store memories at the same time. This means replacing standard computer circuit elements such as transistors, capacitors and inductors with new components called memristors, memcapacitors and meminductors. These components exist right now, in experimental forms, and could soon be combined into a new type of machine called a memcomputer. Memcomputers could have unmatched speed because of their dual abilities. Each part of a memcomputer can help compute the answer to a problem, in a new, more efficient version of today's parallel computing. And because difficult problems are solved by the computer's memory and stored directly into that memory, they will also save all the energy that is currently required to transfer data back and forth within the machine. This brand-new type of computing architecture would change the way computers of all types operate, from the tiny chips in your phone to vast supercomputers. It is, in fact, a design that is close to the way the human brain works, holding memories and processing information in the same neurons.

Complete memcomputers have not yet been built, but our experiments with the components indicate that they could have a huge impact on computer design, global sustainability, power use and our ability to answer vital scientific questions.

• An Electronic, Energy-Efficient Brain

It takes a tiny bit of electricity and a fraction of a second to shuffle data like our Word sentence within a machine. But if you think about what happens when the energy for this back and forth is multiplied across worldwide computing use, it is an enormous operation.

Between 2011 and 2012 power requirements for computer data centers around the globe grew by a staggering 58 percent. Combined, the information and communication sectors now account for approximately 15 percent of global electricity consumption.

We cannot fix it by shrinking transistors to smaller and smaller sizes. The International Technology Roadmap for Semiconductors has forecasted that the transistor industry most likely will hit a technological wall by 2016 because available component materials cannot go down any further in size and maintain their capabilities.

Scientific research on some urgent problems will also hit a wall. Memcomputers, by avoiding the expensive, power-hungry and time-consuming process of constantly transferring data between a CPU and memory, should save a significant amount of energy. They are not, of course, the first information-processing devices to handle calculations and storage in one place. The human brain does this very thing, and memcomputing takes its inspiration from this fast, efficient organ sitting on top of our shoulders.

Traditionally computers have relied on their separation of powers to keep programs and the data they use from interfering with one another during processing. Physical changes in a circuit caused by new data—say, the letters we typed in Word—would change and corrupt the program or the data. This could be avoided if circuit elements in a processor could "remember" the last thing they did, even after the electricity is turned off. Data would still be intact.

• Three Parts of a New Machine

Memcomputing components can do exactly that: process information and store it after the electricity stops. One of these new devices is a memristor. To understand it, imagine a pipe that changes its diameter depending on the direction of water flow.

Now replace the water with electric current and replace the pipe with a memristor. It changes its state depending on the amount of current flowing, as the water pipe changes diameter—a wider pipe has less electrical resistance, and a narrower pipe has more. If you think of resistance as a number and the change in resistance as a process of calculation, a memristor is a circuit element that can process information and then hold it after the current is turned off. Memristors can combine the work of the processing unit and of the memory in one place.

The notion of memristors came from Leon O. Chua, an electrical engineer at the University of California, Berkeley, in the 1970s. Materials used to make circuits did not retain memory of their last state like the imaginary water pipe, so the idea seemed farfetched. But over the decades, engineers and materials scientists were able to exert more and more control over the circuit materials they fabricated, imbuing them with new properties. In 2008 Hewlett-Packard engineer Stanley Williams and his colleagues produced memory elements that could shift resistance and hold their shifted state. They shaped titanium dioxide into an electrical component just tens of nanometers wide. In a paper in *Nature*, the scientists showed that the component retained a state that was determined by the history of current flowing through it. The imaginary pipe was real.

It turns out that these devices can be fabricated with a large variety of materials and can be made just a few nanometers across. Smaller dimensions mean that more of them can be packed into a given area, so they can be crammed into almost any kind of gadget. Many of these components can be made in the same semiconductor facilities we now employ to make computer components and therefore can be fabricated on an industrial scale.

Another key component that could be used in memcomputing is a memcapacitor. Regular capacitors are devices that store electrical charges, but they do not change their state, or capacitance, no matter how many charges are deposited in them. In today's computers they are mainly used in a particular kind of memory, called dynamic random-access memory (DRAM), which stores computer programs in a state of readiness so they can be uploaded quickly to the processor when it calls for them. A memcapacitor, however, not only stores charges but changes its capacitance depending on past voltages applied to it. That gives it both memory and processing ability. Further, because memcapacitors store charges—energy—that power could be recycled during computation, helping to minimize energy consumption by the overall machine (Memristors, in contrast, use all the energy put into them).

The meminductor is the third element of memcomputing. It has two terminals, and it stores energy like a memcapacitor while letting current flow through it like a memristor. Meminductors, too, exist right now. But they are quite large because they rely on big wire magnetic coils, so they would be hard to use in small computers. Advances in materials could change that in the near future, however, as it did for memristors just a few years ago.

In 2010 we started trying to show that memcomputing could handle calculations better than current computer architecture. One problem we focused on was finding a way out of a maze. Devising programs for maze running has long been a way to test the efficiency of computer hardware. Conventional algorithms for solving mazes explore the maze in small consecutive steps. For instance, one of the best-known algorithms is the so-called wall follower. The program traces the wall of the maze through all its twists and turns, avoiding empty spaces where the wall ends, and moves, calculation after painstaking calculation, from the entrance to the exit. This step-by-step approach is slow.

Memcomputing, we have shown in simulations, will solve the maze problem extremely fast. Consider a network of memristors, one at each turn of the maze, all in a state of high resistance. If we apply a single voltage pulse across the entrance and exit points, the current will flow only along the solution path—it will be blocked by dead ends in other paths. As the current flows, it changes the resistances of the corresponding memristors. After the pulse disappears, the maze solution will be stored in the resistances of only those devices that have changed their state. We have computed and stored the solution in only one shot. All the memristors compute the solution in parallel, at the same time.

This kind of parallel processing is completely different from current versions of parallel computing. In a typical parallel machine today, a large number of processors compute different parts of a program and then communicate with one another to come up with the final answer. This still requires a lot of energy and time to transfer information between all these processors and their associated—but physically distinct—memory units. In our memcomputing scheme, it simply is not necessary.

Memcomputing really shows advantages when applied to one of the most difficult types of problems we know of in computer science: calculating all the properties of a large series of integers. This is the kind of challenge a computer faces when trying to decipher complex codes. For instance, give the computer 100 integers and then ask it to find at least one subset that adds up to zero. The computer would have to check all possible subsets and then sum all numbers in each subset. It would plow through each possible combination, one by one, which is an exponentially huge increase in processing time. If checking 10 integers took one second, 100 integers would take 10^{27} seconds—millions of trillions of years.

As with the maze problem, a memcomputer can calculate all subsets and sums in just one step, in true parallel fashion, because it does not have to shuttle them back and forth to a processor in a series of sequential steps. The single-step approach would take just a single second.

Despite these advantages and despite the fact that components have already been made in labs, memcomputing chips are not yet commercially available. At the moment, early versions are being tested in academic facilities and by a few manufacturers to see if these untried designs are robust enough, over repeated use, to replace current memory chips made of standard transistors and capacitors. These chips are the kind you find in USB drives and solid-state memory drives. The tests can take a long time because the components need to last years without failure.

We think some memcomputing designs could be ready for use in the very near future. For instance, in 2013, together with two researchers at the Polytechnic University of Turin in Italy, Fabio Lorenzo Traversa and Fabrizio Bonani, we suggested a concept called dynamic computing random-access memory (DCRAM). The goal is to replace the standard type of memory that, as we have discussed, is used to hold programs and data just before a processor calls for them. In this conventional memory, each bit of information that makes up the program is represented by a charge stored on a single capacitor. That calls for a large number of capacitors to represent one program.

If we replace them with memcapacitors, however, all the different logic operations required by the program can be represented by a much smaller number of memcapacitors in this memory area. Memcapacitors can shift from one logic operation to another almost instantly when we apply different voltages to them. Computing instructions such as "do x AND y", "do x OR y" and "ELSE do z" can be handled by two memcapacitors instead of a large number of fixed regular capacitors and transistors. We do not have to change the basic physical architecture to carry out different functions. In computer terminology, this is called polymorphism, the ability of one element to perform different operations depending on the type of input signal. Our brain possesses this type of polymorphism—we do not need to change its architecture to carry out different tasks—but our current machines do not have it, because the circuits in their processors are fixed. And with memcomputing, of course, because this computation is occurring within a memory area, the time- and power-consuming shuffle back and forth to a separate processor is eliminated, and the result of the program's calculations can be stored in the same place.

These systems can be built with present fabrication facilities. They do not require a major leap in technology. What may hold them up is the need to design new software to control them. We do not yet know the most efficient kinds of operating systems to command these new machines. The machines have to be built, and then various controlling systems have to be tested and optimized. This is the same design process that computer scientists went through with our present crop of machines.

Scientists also would like to find the best way to integrate these new memelements into our current computers. We will need to build, test, rebuild and retest. It is enticing, though, to consider where this technology could lead us. After building and testing, computer users might have a small device, maybe small enough to hold in your hand, that could tackle very complex problems involving, say, pattern recognition or modeling the earth's climate at a very fine scale.

(Excerpt from *Scientific American*, February 2015)

Unit 9
Computer Science

Exercises

Reading Comprehension

Directions: Answer the following questions based on the information from the text.

1. What is a memcomputer? And what influences will it have on computer industry?
2. Why is memcomputer considered an electronic energy-efficient brain?
3. What can a memristor do?
4. What enables a memcapacitor to have both memory and processing ability?
5. In what way is the parallel processing in memcomputer different from the current versions of parallel computing?

Unit 10
Information Technology

导读

本单元的文章均涉及信息技术的应用。Text A详细讲解了最新的信息技术在教育领域的应用，以视频游戏技术为例，描述了人们对视频游戏的不同态度以及游戏在教育中的作用，并详细解释了视频游戏如何应用到教育领域。Text B提供了另一个信息技术应用的具体案例。在这个案例中，作者研发出一套有助于理解医学术语的软件，却意外揭露了某些研究人员剽窃论文和重复申请研究资助计划的违反伦理行为。Text C简短揭秘了各大科技公司引诱顾客加入他们特定数字生态系统的各种花招。

Part I Intensive Reading

Text A

Mind Games

Video games could transform education. But first, game designers, teachers and parents have to move beyond both hype and fear.

By Alan Gershenfeld

In 1993, the year I began my career in video games, the public face of the industry was Mortal Kombat. In this martial-arts fighting game, two players would pummel each other until one opponent was sufficiently stunned—and then deliver a "Fatality" move. One character could grab his opponent's head and then rip his spinal cord out of his still standing body. Not surprisingly, parents, teachers and politicians were horrified. Congress held hearings about the game and its influence on youth. The episode led to the creation of the Entertainment Software Rating Board, which today rates games based on their age appropriateness.

Unit 10
Information Technology

My friends and family thought I was crazy for working in the game industry, particularly because I had left a good career in independent filmmaking to do it. Yet when I started my work as a studio executive at Activision, a popular video game publisher, it quickly became clear that games were much more diverse and textured than most people realized. They were not only an emerging entertainment medium—they were a new art form.

At the core, video games are about verbs, what the player does in a game. These games are about exploring, evaluating, choosing, deciding and solving. For example, Spycraft, an action game we developed with William Colby, former head of the CIA, and Oleg Kalugin, a former major general of the KGB❶, confronted players with complex moral and ethical choices based on real-life experiences. In the simulation game Civilization: Call to Power, players had to make complex decisions about how to build and sustain an empire by balancing cultural, diplomatic, military and scientific advancements.

Although these games had many enthusiastic fans, they were low profile compared with the big action games. By the mid-1990s the public associated video games with first-person-shooter games, in which players careened through three-dimensional environments, mowing down enemies with extravagant weapons. Once it was discovered that the high school shooters in the Columbine massacre of 1999 were avid fans of this genre, video games were again vilified. Today the gap in how video games are perceived is wider than ever. On one hand, conferences, articles and best-selling books are making the case that games and "gamification"—applying the principles of game design to solve real-world challenges—can save the planet. On the other hand, parents struggle with the amount of time their kids spend on digital media—roughly eight hours a day on average.

Yet video games have great potential to help confront the educational challenges of the 21st century. My company, E-Line Media, is working with the National Science Foundation, the Smithsonian Institution, the U.S. Agency for International Development, the Bill & Melinda Gates Foundation, the Mac Arthur Foundation,

---

❶ KGB, abbreviation for Komitet Gosudarstvennoi Bezopasnosti, was the state security police (1954–1991) of the USSR.

the AMD Foundation, the Defense Advanced Research Projects Agency, the White House Office of Science and Technology Policy, Intel, Google, the Massachusetts Institute of Technology's Center for Bits and Atoms, and Arizona State University's Center for Games & Impact, to name a few, all in an effort to figure out how to use video games to improve education. We are learning that it will take a good deal of R&D to get this right.

- **The Class of 2024**

Ten years from now today's second graders will graduate from high school in a world of some eight billion people. They will have to do jobs that do not currently exist, master technologies that have not yet been developed, and build skills that cannot be replaced by technology or outsourced to the cheapest labor. They will need to be scientifically literate and socially adept. They will need to be able to understand complex systems, think critically, propose solutions based on evidence (sometimes emerging and conflicting), and persist despite challenges.

Too many schools do a poor job of fostering these abilities. Most students enter elementary school with a natural curiosity about how the world works, but all too often, by the end of middle school, we have beaten this out of them. Every eight seconds an American public high school student drops out of school; over the next decade that alone will cost the nation an estimated $3 trillion in lost wages, productivity and taxes. Forty-six percent of college students fail to graduate with any credential within six years.

Clearly, for many kids, traditional education is neither relevant nor engaging. Digital games, on the other hand, captivate them. Ninety-seven percent of American teenagers regularly play video games. Fortunately, even games that seem to have no redeeming value can deliver positive, lasting neuropsychological effects. Daphne Bavelier, a psychologist at the University of Geneva, has shown that violent action games can, over time, increase a player's brain plasticity and learning capacity, improve vision and perceptually motivated decision making, sharpen a person's ability to tune out distraction, and strengthen the ability to mentally "rotate" objects.

Games are different from other popular media in that they are interactive and participatory. They enable players to step into different roles (scientist, adventurer, inventor, political leader), confront problems, make choices and explore the consequences. They enable players to advance at their own pace and to fail in a safe environment. Most significant, they give players agency—the ability to make a difference in both virtual and real-world environments.

Scientists are discovering a powerful alignment between good game design and effective learning. This research is emerging at a time of great disruption in education. Low-cost tablets and laptops are becoming ubiquitous in schools, but most teachers are still not sure how to use them in the classroom. Schools nationwide are working to implement the new Common Core Standards and Next Generation Science Standards, which focus on higher-order skills, but traditional curricula and pedagogy are proving ineffective at delivering them.

Game-based learning has the potential to help tackle many of these challenges. Educators can use games to rethink curricula. Students can use them to exercise critical thinking, problem-solving skills, creativity and collaboration. Games can put the joy and wonderment back into science and scientific inquiry.

That is the good news. The bad news is that a large gap exists between the potential and the reality. Most game-based-learning projects have great difficulty making the transition from research into widely used educational products. As a result, the rhetoric around games and learning can feel overhyped. My colleague Michael Angst and I founded E-Line Media to help close this gap. But it will take more than one company. The best game designers in the industry will have to work together with scientists and educators to build games informed by the most recent research into learning, behavior and neuroscience.

- **Games in the Classroom**

Games will have the deepest impact on learning when they become a meaningful part of the school experience. There are a couple of ways this can happen—with "bounded" games that one plays and finishes (a strategy game that can be won, for example) and by using the principles of game design to restructure learning.

New research is enhancing our understanding of both. For example, scientists at the M.I.T. Education Arcade, in collaboration with the developers of a financial-literacy game called Celebrity Calamity, have shown how a bounded game can be a useful precursor to formal learning. The experiment involved two learning sequences: one in which students first played the game and then listened to a lecture and one in which the order was reversed. They found that students who went straight to the lecture did not know what to listen for, whereas students who played the game first had better context and greater motivation.

Teachers who grew up playing games are particularly adept at finding ways to integrate game play into the classroom. As an example, two social studies teachers in Texas, frustrated by their students' hatred of history, developed a middle school history curriculum inspired by the commercial video game Civilization. They called it Historia. Working on paper, teams of students led fictional civilizations, competing alongside (and sometimes against) the great empires of the past. Students researched history to understand how their decisions would impact the economic, military and cultural strength of their civilization. Initially the teachers encountered resistance from parents and administrators, but once standardized test scores started improving, the dissent quickly disappeared. At E-Line, we are now working on a digital version of Historia, which we will pilot this spring and release this fall.

As it turns out, making a good video game also requires a complex set of higher-order skills—thinking analytically and holistically, experimenting with and testing out theories, creating and collaborating with peers and mentors. That is why the M.I.T. Media Lab developed a programming language, Scratch, that enables kids as young as kindergartners to build games. Microsoft has developed a similar tool called Kodu. And high schools and colleges are increasingly offering instruction in tools used in professional game creation such as Unity, Flash and Java.

At E-Line, our contribution to this genre is Gamestar Mechanic, which we are developing in partnership with the MacArthur Foundation and the New York City-based nonprofit Institute of Play. The game is designed for students between the ages of eight and 14. Working solo or in groups, they log on to a PC or Mac and learn the fundamentals of game design by playing and fixing broken games. On a community site, they can publish and collaborate on games. They can review games, reflect on their own ideas and defend their design decisions. Since Gamestar Mechanic was launched in the fall of 2010, more than 6,000 schools and after-school programs have started using it. Students have published more than 500,000 original games, which have been played more than 15 million times in 100 countries.

Game designers are also adapting commercial games for the classroom. SimCityEdu, for example, is an educational version of the famous simulation game SimCity, created through a partnership among the Bill & Melinda Gates Foundation, the MacArthur Foundation, the game company Electronic Arts, the Entertainment Software Association, the Institute of Play, the publisher Pearson and the Educational Testing Service, which administers the SAT. The company Valve has also developed an educational version of its popular game Portal, in which the player is dropped into a mysterious laboratory and has to solve a series of puzzles to survive. The educational version, called Teach with Portals, is designed to make "physics, math, logic, spatial reasoning, probability, and problem-solving interesting, cool and fun".

- **Education by Stealth**

Kids are unlikely to embrace Call of Duty: Calculus in their discretionary playtime. Nevertheless, we believe there is a large audience for games that explore challenging themes and that open new worlds—as long as they are truly great games.

There are precedents in other media. In the film industry, for example, Participant Media has had success making movies that "inspire and accelerate social change". Examples include *Good Night, and Good Luck*, *Syriana* and *Lincoln*.

We think this same approach can work with games. Many game designers have families of their own and would rather use their craft to empower youth than to work on yet another $50-million first-person shooter. At E-Line, our first major project in this field is a collaboration with the Cook Inlet Tribal Council (CITC), a pioneering Alaska Native social service organization. CITC has launched the first U.S.-based indigenous-owned video game company, Upper One Games. Together we are developing a new genre—game-based cultural storytelling—that emphasizes cultural heritage and intergenerational wisdom. The first consumer game we will release is the action-adventure game Never Alone, in which the player will take on the role of a young Inupiat girl facing a struggle for survival. Along with her companion, a young fox, the player must overcome obstacles and fears in the harsh and beautiful Arctic landscape. The game is framed as a series of interconnected stories told by elders to youth; both the narrative and core game-play mechanics explore how interdependence, adaptation and resiliency are critical for survival in challenging circumstances. It will be available for game consoles (Sony PlayStation and Microsoft Xbox) as well as for PCs and Macs.

So far, though, the best example of a game that transcends commercial and educational boundaries has to be Minecraft. It is a phenomenon unlike any that I have seen in my career. Originally developed by Swedish programmer Markus "Notch" Persson, the game has become a global phenomenon, with more than 25 million players, mostly tweens. Minecraft players roam freely and build Lego-like worlds, either individually or collaboratively. In "Survival" mode, the player must build shelter before it gets dark and the bad guys come out. To do so, the player must find the resources ("mine") and make tools ("craft"). Once safe from the bad guys—or in the game's enemy-free "Creative" mode—players build almost anything. A quick whirl through Minecraft creations on YouTube will reveal models of virtually every iconic building on the globe—the Eiffel Tower, the Taj Mahal and, my favorite, a scale model of China's Forbidden City, built from nearly 4.5 million blocks, complete with a roller coaster to take you on a tour.

Not only is Minecraft immersive and creative, it is also an excellent platform for making almost any subject area more engaging. We recently worked on a project with Google, the California Institute of Technology, TeacherGaming (co-founded by Joel Levin, a private school teacher in New York City who began using Minecraft in the classroom shortly after the game was released and quickly gained a global following as the "Minecraft Teacher") and leading Minecraft "modder" Daniel Ratcliffe to develop qCraft, a modification ("mod") to the game that introduces players to the bizarre world of quantum mechanics.

To demonstrate the concept of observational dependency, qCraft blocks change shape and color depending on who is looking at them and from which direction. Entangled blocks are inextricably linked, even if they are a vast distance apart. Superpositional blocks are more than one thing at once.

On the qCraft blog last November, Levin explained the rationale for the project: "By the time our 7-year-old finishes grad school, quantum computers may be commonplace...Some of the hardest problems in medicine, aerospace, statistics, and more will be tackled by machines using qubits instead of bits.... It is our firm belief that when a young person who has played qCraft encounters these challenging concepts again, they will have an increased intuitive understanding."

- **The Next Move**

Realizing the full educational potential of games will involve addressing the good and the bad. Many parents, teachers and policy makers are still skeptical.

An ongoing concern is violence—the question of whether playing violent video games leads to real-world violent behavior. The issue is highly polarized. The game industry points to countries such as Japan and South Korea, avid consumers of violent games that also have some of the lowest rates of gun violence in the world. They also highlight multiple studies showing that while playing violent games may increase short-term aggressive behavior, there is no correlation to the type of violent behavior exhibited by, for example, school shooters. On the other side of the debate, many parents will refer to a cluster of studies that reinforce some of the connections between games and violence. They will argue that because games can have positive learning effects, does it not stand to reason that they can have negative effects as well?

An increasing number of parents also express concern about the amount of time their children spend playing games. Digital media consumption is like food consumption—it is important to have a balanced diet, and each person's diet is different. The more informed and engaged the parent, the better the outcome for the child. By playing games with their kids, parents can become more savvy observers. Innovative approaches to game design can also help. Games can be optimized for shorter play cycles, or they can incorporate real-world activities—exercise tracked through an accelerometer, for example—into game-play loops.

Technology and design advances will make video games ever more realistic, fantastical and ubiquitous. We will see gaming extend into consumer virtual-reality devices, wearable computing, and beyond. These new technologies will unlock opportunities to use games for social good. They are also likely to intensify the concerns that parents and policy makers already have. That is why it is so important that, starting now, we give video games the proper attention they deserve.

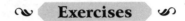
(Excerpt from *Scientific American*, February 2014)

∞ Exercises ∽

I. Reading Comprehension

• **Section One**

Directions: Answer the following questions based on the information from the text.

1. Why did the author's friends and family think he was crazy for working in the game industry?
2. Why does the author say today the gap in how video games are perceived is wider than ever?
3. What roles do games play in education according to the author?
4. Why is a bounded game a useful precursor to formal learning? Illustrate the point with examples.
5. What are people's attitudes towards violence in games?

• **Section Two**

Directions: Write an abstract based on the text in no more than 200 words.

Abstract:

Key words:

Unit 10
Information Technology

II. Vocabulary

• Section One

Directions: Choose the word or phrase that is closest in meaning to the underlined part in each sentence.

1. The episode led to the creation of the Entertainment Software Rating Board, which today <u>rates</u> games based on their age appropriateness.
   A. assigns or values
   B. ranks
   C. is worth of
   D. has a high opinion of
2. Forty-six percent of college students fail to graduate with any <u>credential</u> within six years.
   A. qualification   B. achievement   C. certificate   D. quality
3. Once it was discovered that the high school shooters in the Columbine massacre of 1999 were <u>avid</u> fans of this genre, video games were again vilified.
   A. keen   B. anxious   C. greedy   D. hungry
4. Clearly, for many kids, traditional education is neither relevant nor <u>engaging</u>.
   A. charming
   B. busy
   C. agreeing to marry
   D. involving
5. This research is emerging at a time of great <u>disruption</u> in education.
   A. interruption
   B. disturbance
   C. innovation
   D. disorderly outburst
6. Kids are unlikely to embrace Call of Duty: Calculus in their <u>discretionary</u> playtime.
   A. available for use at the freedom of the user
   B. (especially of funds) not earmarked
   C. careful
   D. cautious
7. Entangled blocks are <u>inextricably</u> linked, even if they are a vast distance apart.
   A. confusedly
   B. impossible to escape from
   C. unsolvable
   D. complicatedly
8. Scientists are discovering a powerful <u>alignment</u> between good game design and effective learning.
   A. arrangement in a straight line
   B. a position of agreement or alliance
   C. the route of a railway or road

D. apparent meeting of two or more celestial bodies in the same degree of the zodiac

9. Realizing the full educational potential of games will involve addressing the good and the bad.

   A. speaking to

   B. saying or writing remarks or a protest to

   C. writing the name and address

   D. thinking about and beginning to deal with

10. They also highlight multiple studies showing that while playing violent games may increase short-term aggressive behavior, there is no correlation to the type of violent behavior exhibited by, for example, school shooters.

    A. ready to attack   B. fast-growing

    C. pursuing one's aims and interests forcefully   D. having pioneering spirit

● **Section Two**

Directions: There are two or three meanings for each semi-technical term underlined in the following sentences. Choose the correct one according to the context.

1. Congress held hearings (the faculties of perceiving sounds; opportunities to state one's case; acts of listening to evidence in a court of law or before an official) about the game and its influence on youth.

2. By the mid-1990s the public associated video games with first-person-shooter games, in which players careened through 3D environments, mowing (cutting down an area of grass with a machine; killing someone with a fusillade of bullets or other missiles) down enemies with extravagant weapons.

3. Although these games had many enthusiastic fans, they were low profile (an outline of something, especially a person's face, as seen from one side; an outline of part of the earth's surface; the extent to which a person or organization that attracts public notice or comment) compared with the big action games.

4. Daphne Bavelier has shown that violent action games can, over time, increase a player's brain plasticity (the quality of being easily shaped or moulded; the adaptability of an organism to changes in its environment or differences between its various habitats), and learning capacity.

Unit 10
Information Technology

5. Most significant, they give players <u>agency</u> (a business that provides a service on behalf of other businesses; government organization responsible for a certain area of administration; action or intervention, especially such as to produce a particular effect).

6. There are a couple of ways this can happen—with "<u>bounded</u>" (walked or ran with leaping strides; formed the boundary of or enclosed) games that one plays and finishes and by using the principles of game design to restructure learning.

7. At E-Line, we are now working on a digital version of Historia, which we will <u>pilot</u> (act as a pilot of an aircraft or ship; guide or steer; test a scheme, project etc. before introducing it more widely) this spring and release this fall.

8. The first consumer game we will <u>release</u> (set free; allow something concentrated in a small area to spread and work freely; make available for general viewing or purchase) is the action-adventure game Never Alone.

9. To do so, the player must find the resources ("<u>mine</u>") (dig in the earth for coal or other minerals; delve into an abundant source to extract something of value, especially information or skill; lay explosives just below the surface of ground) and make tools.

10. They can tell whether their child is learning to <u>code</u> (convert a message into a particular system of symbols for the purposes of secrecy; write program instructions; be the genetic determiner of a characteristic) in Minecraft or playing a 50th Hunger Games death match.

- **Section Three**

Directions: Match the Chinese terms with their English equivalents.

1. 娱乐媒体   A. video game
2. 高阶技能   B. learning ability
3. 游戏化学习   C. gamification
4. 教育学   D. Entertainment Software Rating Board
5. 金融知识   E. entertainment medium
6. 游戏化   F. redeeming value
7. 娱乐软件评级委员会 G. pedagogy
8. 学习能力   H. higher-order skill
9. 视频游戏   I. game-based learning

10. 可取价值　　　　　　　　J. curriculum
11. 空间推理力　　　　　　　K. financial literacy
12. 课程　　　　　　　　　　L. qubit
13. 游戏控制台　　　　　　　M. quantum mechanics
14. 量子比特　　　　　　　　N. spatial reasoning
15. 量子力学　　　　　　　　O. game console

III. Questions for Discussion

Directions: Work in groups and discuss the following questions.

1. How do on-line games influence players?
2. What qualities should one have if he or she wants to design a game for educational purpose?
3. What is the good and bad of game-based learning?

Unit 10
Information Technology

Part II Extensive Reading

Text B

The Case of Stolen Words

The author wanted to build software that would navigate medical jargon. He ended up uncovering widespread plagiarism and hundreds of millions of dollars in potential fraud.

By Harold "Skip" Garner

In 1994 I reinvented myself. A physicist and engineer at General Atomics, I was part of an internal think tank charged with answering hard questions from any part of the company. Over the years, I worked on projects as diverse as cold fusion and Predator drones. But by the early 1990s I was collaborating frequently with biologists and geneticists. They would tell me what cool new technologies they needed to do their research; I would try to invent them.

Around that time I heard about a new effort called the Human Genome Project. The goal was to decipher the sequence of the approximately three billion DNA bases, or code letters, in human chromosomes. And before I knew it, I found myself a professor at the University of Texas Southwestern Medical Center, where my scientific partner, a geneticist, and I were building one of the Human Genome Project's first research centers.

Everything was different there. My colleagues spoke a different language—medicine. I spoke physics. In physics, basic equations govern most everything. In medicine, there are no universal equations—just many observations, some piecewise understanding and a tremendous amount of jargon. I would attend seminars and write down huge lists of words I had never heard and then spend hours afterward looking them up. To read a scientific paper, I had to have a medical dictionary on hand.

Frustrated with my inability to understand any contiguous piece of text, I decided to develop software to help me. I wanted a search engine that would take a chunk of text and return references for further reading, abstracts and papers that would quickly get me up to speed on the topic at hand. It was a tough problem. Search engines for the Web were just emerging. They were fine for finding the best falafel restaurant in town, but they could not begin to digest a paragraph containing multiple interrelated concepts and point me to related readings.

With some students and postdocs, I set about studying text analytics, and together we developed a piece of software called eTBLAST (electronic Text Basic Local Alignment Search Tool). It was inspired by the software tool BLAST, used to search DNA and protein sequence databases. A query for BLAST was usually a series of 100 to 400 DNA letters and would return longer sequences that included those codes. The query for eTBLAST would be a paragraph or page—typically 100 words or more. Designing the search protocol was harder than designing software to seek a string of letters because the search engine could not merely be literal. It also had to recognize synonyms, acronyms and related ideas expressed in different words, and it had to take word order into account. In response to a query consisting of a chunk of text, eTBLAST would return a ranked list of "hits" from the database it was searching, along with a measure of the similarity between the query and each abstract found.

The obvious database to search was Medline, the repository, maintained by the National Library of Medicine at the National Institutes of Health, of all biological research relevant to medicine. It contains the title and abstract of millions of research papers from thousands of peer-reviewed journals. Medline had a search engine that was keyword-based, so a query of a few words—for example, "breast cancer genes"—would return plenty of hits, often with links to full papers. But as a newly converted biomedical researcher, I did not even know how to start many of my searches.

The first versions of eTBLAST took hours to compare a paragraph of a few hundred words against Medline. Using eTBLAST, I could make my way through scientific papers, mastering their meaning paragraph by paragraph. I could pop a graduate student's thesis proposal in and quickly get up to speed on the pertinent literature. Then events took a strange turn. A couple of times I found text in student proposals that was identical to text in other, uncited papers. The students received remedial ethics training. I received a research question that would change my career: How much of the professional biomedical literature was plagiarized?

- **Déjà Vu[1]**

When I set out to explore this new question, the research on plagiarism in biomedicine consisted of anonymous surveys. In the most current survey I found, researchers admitted to plagiarizing 1.4 percent of the time. But the accuracy of that number depended on the honesty of the survey respondents. With eTBLAST, we could find out whether they were telling the truth.

Once we had enough student help and a sufficiently powerful computer, we randomly selected abstracts from Medline and then used them as eTBLAST queries. The computer would compare the query text with the entire contents of Medline, looking for similarities, then return a list of hits. Each hit came with a similarity score. The query was always at the top of the list—100 percent similarity. The second hit typically had a similarity score between the single digits and 30 percent. Occasionally, though, we found that the second and sometimes third hits had scores close to 100 percent. After running a few thousand queries, we started to see that about 5 percent of queries had suspiciously high similarity scores. We reviewed those abstracts by eye to make sure the software was finding things that a human would consider similar. Then we went on to compare the full text of papers that had suspiciously similar abstracts. Soon we began to find blatant examples of plagiarism—not just recycled phrases but entire papers lifted whole cloth. It was disappointing, even astounding. It is quite a different thing to see examples of plagiarized papers side by side.

The next step was to scale up the computing and the analysis. To be thorough, we wanted to perform similarity searching on every entry of sufficient length in Medline—at the time, almost nine million entries, each containing an average of 300 words, times nearly nine million comparisons. The task took months and consumed a considerable amount of our lab's computing power. As the results emerged, we

---

[1] Déjà Vu means the feeling that you have already experienced the things that are happening to you now.

analyzed them and placed all the highly similar results in a database we called Déjà Vu. Déjà Vu began to fill with pairs of highly similar Medline abstracts—about 80,000 pairs that were at least 56 percent similar. The vast majority of these pairs were highly similar for perfectly good reasons—they were updates to older papers, or meeting summaries, for example. But others were suspicious.

We submitted a paper to *Nature* that contained data on the frequency of plagiarism and duplicate publication (sometimes called self-plagiarism), details on the content of the Déjà Vu database and some prime examples. The editors accepted, but because we referred to some abstracts as plagiarized, the lawyers ripped the paper apart. They had an excellent point: The only people who could make a plagiarism determination were editors and ethics review boards. We could present only facts—the amount of text overlap or similarity between any two pieces of scientific literature.

When the *Nature* report came out, all hell broke loose. Journal editors were upset because it gave them extra work to do. To protect their copyright, the editors of the original papers had to insist that the plagiarized papers be retracted. The second publisher of course, was embarrassed. Scientists were angry because our results seemed to expose a flaw in peer review. But everyone grudgingly admitted that this was an important topic and a serious problem. Scientists and clinicians make critical decisions based on what they read in the literature. What did it mean if those decisions were based on tainted studies?

Ultimately we determined that 0.1 percent of professional publications were blatantly plagiarized from the work of others. Some 1 percent were self-plagiarized; one author's work would appear, often nearly verbatim, in as many as five journals. If these percentages seem small, consider that some 600,000 new biomedical papers are published every year. And before long, we noticed that the publishing process had begun to change. Journal editors started using eTBLAST to check their submissions. I had changed, too. I had evolved again, adding "ethics researcher" to my job description.

Unit 10
Information Technology

• My Life as an Ethics Cop

The first big plagiarism study was just the beginning. Understanding the causes of plagiarism and their effects on science would require much more work. When is repeated text acceptable? When and why do scientists plagiarize? What other kinds of unethical behavior could textual analysis uncover? So we refined our software, expanded our databases and took on new studies.

Some of our subsequent work revealed unexpected nuances in the plagiarism debate. We found that in some cases, textual similarity is not only acceptable but preferred. In the methods section of a research paper, for example, where the most important consideration is reproducibility of results, unoriginal phrasing serves the important purpose of showing clearly that exactly the same protocol was used.

We also found some truly egregious ethical lapses. In a study published in *Science*, we took the most blatant examples of plagiarism we could find—pairs of papers in which paper B was on average 86 percent identical to paper A—and analyzed them in detail. We e-mailed annotated copies of the papers, along with confidential surveys, to the authors and editors involved with those papers. Ninety percent of the people we contacted responded.

Some of the authors divulged striking ethics violations. Some admitted that they had copied papers while they were reviewing them—and that they had given those papers bad reviews to block their publication. Others blamed the lapse on fictitious medical students. Unsurprisingly, most of the tainted papers in that bunch have since been retracted.

These were not the last ethics violations we would find. In early 2012 we began looking for instances of double-dipping on grants—that is, getting money from multiple government agencies to do the same work. We downloaded summaries of approximately 860,000 grants from government and private agencies, including the National Institutes of Health, the National Science Foundation, the Department of Defense, the Department of Energy and Susan G. Komen for the Cure, and subjected them to the eTBLAST treatment. The study required 800,000 times 800,000 (roughly 1012) comparisons and supercomputer-level power.

After reviewing the 1,600 most similar grant summaries, we found that about

170 pairs had virtually identical goals, aims or hypotheses. We concluded several things: That double-dipping had been happening consistently for a long time; that it involved America's most prestigious universities; and that the resulting loss to biomedical research was as high as $200 million a year.

● The Future of Scientific Publishing

A small percentage of people have always broken societal norms, and scientists are no different. Text analytics gives us a good tool for policing bad behavior. But it could eventually do much more than smoke out plagiarism. It could facilitate entirely new ways of sharing research.

One intriguing idea is to adopt a Wikipedia model: to create a dynamic, electronic corpus of work on a subject that scientists continually edit and improve. Each new "publication" would consist of a contribution to the single growing body of knowledge; those redundant methods sections would become unnecessary. The Wikipedia model would be a step toward a central database of all scientific publications across all disciplines. Authors and editors could use text mining to verify the novelty of new research and to develop reliable metrics for the impact of an idea or discovery. Ideally, instead of measuring a paper's impact by the number of citations it receives, we would measure its influence on our total scientific knowledge and even on society.

At Virginia Tech, we are struggling to keep eTBLAST running, but the software still has thousands of users. We are working to apply the kind of paragraph-size similarity searching that uncovered so many instances of plagiarism to other ends, including grant management, market research and patent due diligence.

(Excerpt from *Scientific American*, March 2014)

Unit 10
Information Technology

Exercises

I. Translate the following technical terms into English.

1. 智库 _____
2. 自我剽窃 _____
3. 冷聚变 _____
4. 无人机 _____
5. 染色体 _____
6. DNA碱基对 _____
7. 医学术语 _____
8. 计算能力 _____
9. 文本分析 _____
10. 搜索协议 _____
11. 查询 _____
12. 同行评议 _____
13. 基本方程 _____
14. 生物医学文献 _____
15. 匹配得分 _____
16. 伦理审查 _____
17. 结果的可再现性 _____
18. 动态电子语料库 _____
19. 开题报告 _____
20. 专利尽职调查 _____

II. Translate the following paragraphs into Chinese.

The obvious database to search was Medline (available from PubMed at pubmed.org), the repository, maintained by the National Library of Medicine at the National Institutes of Health, of all biological research relevant to medicine. It contains the title and abstract of millions of research papers from thousands of peer-reviewed journals. Medline had a search engine that was keyword-based, so a query of a few words—for example, "breast cancer genes"—would return plenty of hits, often with links to full papers. But as a newly converted biomedical researcher, I did not even know how to start many of my searches.

The first versions of eTBLAST took hours to compare a paragraph of a few hundred words against Medline. But the software worked. Using eTBLAST, I could make my way through scientific papers, mastering their meaning paragraph by paragraph. I could pop a graduate student's thesis proposal in and quickly get up to speed on the pertinent literature. My research partners and I even spoke with Google about commercializing our software, only to be told it did not fit with the company's business model.

Then events took a strange turn. A couple of times I found text in student proposals that was identical to text in other, uncited papers. The students received remedial ethics training. I received a research question that would change my career: How much of the professional biomedical literature was plagiarized?

Text C

The Tech Jungle

How big tech companies lure you into their digital ecosystems.

By David Pogue

The question is no longer, "What phone should I get?"

That was an important question immediately after the arrival of the iPhone and its competitors. But now it's time to admit that today's smartphones (and tablets) are nearly identical. Apple and Google (maker of Android phone software) have copied each other's ideas so completely that the resultant phones are incredibly close in looks, price, speed and features.

These days the Apples, Googles and Microsofts of the world are competing on a different battlefield: They're racing to build the best, most enticing ecosystem. Each is creating a huge archipelago of interconnected products and services. It's about velvet handcuffs: making it easy for you to embrace its offerings and as hard as possible to switch to a rival's.

A typical ecosystem includes hardware (phone, tablet, laptop, smart-watch, TV box); online stores (music, movies, TV, e-books); synchronization of your data across gadgets (calendar, bookmarks, notes, photographs); cloud storage (a free online "hard drive" for files); and payment systems (wave your watch or phone instead of swiping a credit card).

For consumers, the choice is now what suite of products they like best.

If you're one of these companies, though, you've got a difficult decision to make: Should you open up your services to people who use your competitors' products?

Say, let an iPhone user load an Outlook calendar or let a Microsoft Band smartwatch wearer sync data to an Android tablet.

On one hand, making your software available to those outside your ecosystem could introduce the rest of the world to the superiority of your products—and possibly bring in new consumers. On the other hand, you would lose the exclusivity of those services as a lure. Why would anyone switch if she or he can already get the best of a rival's offerings?

So what approach are the giants taking? It's a mixed bag.

Apple is the most closed. In general, it writes apps only for iPhones and iPads. You can't make a FaceTime call to an Android or Windows Phone, for example, or run the Apple Maps app on those devices (not that you'd want to). And you can't use the Apple Watch with anything but an iPhone. You can, however, use Apple's iCloud (online file storing and sync services) on a Windows device—but not on one using Google's Android.

Google goes to great lengths to make its wares available to other platforms. If you have an iPhone, you can use Google's apps (Gmail, Chrome, Google Maps), services (Docs, Sheets, Slides) and even digital store (Books, Music, Newsstand). The services and store are also available to Mac, Windows and Linux users. You can even link an Android Wear smart watch with an iPhone. Then there's Microsoft. Microsoft Office is available for just about anything with a screen, as are many of its mobile apps.

Why such inconsistency?

It helps to understand the individual corporate motives. Although these three companies offer so many similar (okay, almost identical) gadgets and services, each is actually running on an entirely different business model. Apple is primarily in the business of selling hardware; Microsoft, software; Google, ads. Each has different considerations in calculating what to open up.

And Apple and Google continue to branch out; both now offer, if you can believe it, software for your car dashboard and home-automation system designed to work with their respective smart phones. Surely Microsoft won't be far behind. Samsung boasts its own cluster of competitive products and linked services. Even Amazon—

once a bookstore, for goodness's sake—now makes phones, tablets and TV boxes.

You, the consumer, should be delighted by this direction. Perhaps dismayed by all the duplication of effort but happy there's competition, which always begets innovation (and often lower prices). And you should be pleased that overall the trend seems to be for these companies to make more of their services accessible, no matter which phone or computer you own.

Eventually the ecosystems may well become nearly identical, too. Maybe at that point, the question will once again become, "What phone should I get?"

(Excerpt from *Scientific American*, December 2015)

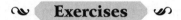

Reading Comprehension

Directions: Answer the following questions based on the information from the text.

1. What does the title "The Tech Jungle" mean?
2. Is the question "What phone should I get" answered at the beginning and end of the reading? Why?
3. Where are the giants competing with each other? Why do they do so?
4. What does a typical digital ecosystem mentioned in the article refer to? What influence does it have on a consumer?
5. What is the trend of the digital ecosystem?

Unit 11
Cyber Security

> **导读**
>
> 本单元的文章均涉及网络安全问题。Text A主要讲述了美国国家安全局经历斯诺登泄密事件后的困境，并从信息存储安全的角度详细阐述了大数据时代政府及相关部门应该如何预防网络信息安全问题。Text B阐述了网络攻击的现状，并从不同角度提出相应的解决方法。Text C简短论述了高科技公司如何从众星捧月般的地位跌落至失去顾客信任的地步。

Part I   Intensive Reading

Text A

Saving Big Data from Itself

A three-step plan for using data right in an age of government overreaching.

By Alex "Sandy" Pentland

For the first few decades of its existence, the National Security Agency was a quiet department with one primary job: keeping an eye on the Soviet Union. Its enemy was well defined and monolithic. Its principal tools were phone taps, spy planes and hidden microphones.

After the attacks of September 11, all of that changed. The NSA's chief enemy became a diffuse network of individual terrorists. Anyone in the world could be a legitimate target for spying. The nature of spying itself changed as new digital communication channels proliferated. The exponential growth of Internet-connected mobile devices was just beginning. The NSA's old tools apparently no longer seemed sufficient.

In response, the agency adopted a new strategy: collect everything. As former NSA director Keith Alexander once put it, when you are looking for a needle in a haystack❶, you need the whole haystack. The NSA began collecting bulk phone call records from virtually every person in the U.S.; soon it was gathering data on bulk Internet traffic from virtually everyone outside of the U.S. Before long, the NSA was collecting an amount of data every two hours equivalent to the U.S. Census.

The natural place for the NSA to store this immense new haystack was the same place it had always stored intelligence assets: in the agency's own secure facilities. Yet such concentration of data had consequences. The private, personal information of nearly all people worldwide was suddenly a keystroke away from any NSA analyst who cared to look. Data hoarding also made the NSA more vulnerable than ever to leaks. Outraged by the scope of the NSA's secret data-collection activities, then NSA contractor Edward Snowden managed to download thousands of secret files from a server in Hawaii, hop on a flight to Hong Kong and hand the documents over to the press.

Data about human behavior have always been essential for both government and industry to function. But a secretive agency collecting data on entire populations, storing those data in clandestine server farms and operating on them with little or no oversight is qualitatively different from anything that has come before. No surprise, then, that Snowden's disclosures ignited such a furious public debate.

So far much of the commentary on the NSA's data-collection activities has focused on the moral and political dimensions. Less attention has been paid to the structural and technical aspects of the NSA debacle❷. Not only are government policies for collecting and using big data inadequate, but the process of making and evaluating those policies also needs to move faster. Government practices must adapt as quickly as the technology evolves. There is no simple answer, but a few basic principles will get us on track.

---

❶ A needle in a haystack means something that is almost impossible to find because it is hidden among so many other things.
❷ A debacle is an event or attempt that is a complete failure.

Unit 11
Cyber Security

- **Step 1   Scatter the Haystack**

   Alexander was wrong about searching for needles in haystacks. You do not need the entire stack—only the ability to examine any part of it. For governments, it makes devastating leaks that much more likely. For individuals, it creates the potential for unprecedented violations of privacy.

   The Snowden disclosures made clear that in government hands, information has become far too concentrated. The NSA and other government organizations should leave big data resources in place, overseen by the organization that created the database, with different encryption schemes. Different kinds of data should be stored separately: financial data in one physical database, health records in another, and so on. Information about individuals should be stored and overseen separately from other sorts of information. The NSA or any other entity that has good, legal reason to do so will still be able to examine any part of this far-flung haystack. It simply will not hold the entire stack in a single server farm.

   The easiest way to accomplish this disaggregation is to stop the hoarding. Let the telecoms and Internet companies retain their records. There need be no rush to destroy the NSA's current data stores, because both the content of those records and the software associated with them will quickly become ancient history.

   It might be hard to imagine the NSA giving up its data-collection activities, but doing so would be in the agency's own interest. The NSA seems to know this, too. Speaking at the Aspen Security Forum in Colorado last summer, Ashton B. Carter, then deputy secretary of defense, diagnosed the source of the NSA's troubles. The "failure (of the Snowden leaks) originated from two practices that we need to reverse...There was an enormous amount of information concentrated in one place. That's a mistake." And second, "you had an individual who was given very substantial authority to access that information and move that information. That ought not to be the case, either." Distributed, encrypted databases running on different computer systems would not only make a Snowden-style leak more difficult but would also protect against cyberattacks from the outside. Any single exploit would likely result in access to only a limited part of the entire database.

   How does distributing data help protect individual privacy? The answer is that it makes it possible to track patterns of communication between databases and human operators. Each category of data-analysis operation, whether it is searching for a particular item or computing some statistic, has its own characteristic pattern of communication—its own signature web of links and transmissions among databases. These signatures, metadata about metadata, can be used to keep an eye on the overall patterns of otherwise private communications.

Consider an analogy: When patterns of communication among different departments in a company are visible (as with physical mail), then the patterns of normal operations are visible to employees even though the content of the operations (the content of the pieces of mail) remains hidden. In the same way, structuring big data operations so that they generate metadata about metadata makes oversight possible. Telecommunications companies can track what is happening to them. Independent civic entities, as well as the press, could use such data to serve as an NSA watchdog. With metadata about metadata, we can do to the NSA what the NSA does to everyone else.

● **Step 2 Harden Our Transmission Lines**

Eliminating the NSA's massive data stores is only one step toward guaranteeing privacy in a data-rich world. Safeguarding the transmission and storage of our information through encryption is perhaps just as important. Without such safeguards, data can be siphoned off without anyone knowing. This form of protection is particularly urgent in a world with increasing levels of cybercrime and threats of cyberwar.

Everyone who uses personal data, be they a government, a private entity or an individual, should follow a few basic security rules. External data sharing should take place only between data systems that have similar security standards. Every data operation should require a reliable chain of identity credentials so we can know where the data come from and where they go. All entities should be subject to metadata monitoring and investigative auditing, similar to how credit cards are monitored for fraud today.

A good model is what is called a trust network. Trust networks combine a computer network that keeps track of user permissions for each piece of data within a legal framework that specifies what can and cannot be done with the data—and what happens if there is a violation of the permissions. By maintaining a tamper-proof history of provenance and permissions, trust networks can be automatically audited to ensure that data-usage agreements are being honored.

Long-standing versions of trust networks have proved to be both secure and robust. The best known is the Society for Worldwide Interbank Financial Telecommunication (SWIFT) network, which some 10,000 banks and other organizations use to transfer money. SWIFT's most distinguishing feature is that it has never been hacked (as far as we know). Trillions of dollars move through the network every day. Because of its built-in metadata monitoring, automated auditing systems and joint liability, this trust network has not only kept the robbers away, it has also made sure the money reliably goes where it is supposed to go.

Trust networks used to be complex and expensive to run, but the decreasing cost of computing power has brought them within the reach of smaller organizations and even individuals. As the use of trust networks becomes more widespread, it will become safer for individuals and organizations to transmit data among themselves, making it that much easier to implement secure, distributed data-storage architectures that protect both individuals and organizations from the misuse of big data.

- **Step 3 Never Stop Experimenting**

The final and perhaps most important step is for us to admit that we do not have all the answers, and, indeed, there are no final answers. As technology changes, so must our regulatory structures. This digital era is something entirely new; we cannot only rely on existing policy or tradition. Instead we must constantly try new ideas in the real world to see what works and what does not.

Pressure from other countries, citizens and tech companies has already caused the White House to propose some limits on NSA surveillance. Tech companies are suing for the right to release information about requests from the NSA in an effort to restore trust. And in May the House of Representatives passed the U.S.A. Freedom Act; though considered weak by many privacy advocates, the bill would begin to restrict bulk data collection and introduce some transparency into the process.

Those are all steps in the right direction. Yet any changes we make right now will only be a short-term fix for a long-term problem. Ultimately, the most important change that we could make is to continuously experiment and to conduct small-scale tests and project deployments to figure out what works, keep what does and throw out what does not.

(Excerpt from *Scientific American*, August 2014)

Exercises

I. Reading Comprehension

● **Section One**

Directions: Answer the following questions based on the information from the text.

1. What was NSA's chief job before and after the attacks of September 11?
2. What is the consequence of Edward Snowden's disclosure?
3. What is the most important in information security?
4. In what way are transmission lines hardened?
5. According to the author, what should government organizations do with information security?

● **Section Two**

Directions: Write an abstract based on the text in no more than 200 words.

Abstract:

Key words:

Unit 11
Cyber Security

II. Vocabulary

• Section One

Directions: Choose the word or phrase that is closest in meaning to the underlined part in each sentence.

1. The nature of spying itself changed as new digital communication channels proliferated.
   A. reproduced rapidly
   B. reproduced something rapidly in large quantities
   C. increased rapidly in numbers
   D. divided quickly

2. Data hoarding also made the NSA more vulnerable than ever to leaks.
   A. susceptible to attack
   B. susceptible to criticism or persuasion or temptation
   C. capable of being wounded or hurt
   D. liable to higher penalties

3. No surprise, then, that Snowden's disclosures ignited such a furious public debate.
   A. caught fire   B. aroused (an emotion)
   C. inflamed   D. instigated

4. So far much of the commentary on the NSA's data-collection activities has focused on the moral and political dimensions.
   A. aspects or features of a situation
   B. measurable extents of some kind
   C. modes of linear extension
   D. expressions for a derived physical quantity

5. Information about individuals should be stored and overseen separately from other sorts of information.
   A. supervised   B. overlooked
   C. peeped   D. spied

6. Even authoritarian governments should have an interest in distributing data: Concentrated data could make it easier for insiders to stage a coup.
   A. diagnose or classify

B. cause something unexpected to happen

C. organize and participate in

D. present a performance

7. By maintaining a tamper-proof history of <u>provenance</u> and permissions, trust networks can be automatically audited to ensure that data-usage agreements are being honored.

   A. browsing behavior

   B. real name authentication

   C. a record of ownership

   D. place of origin or beginning of something's existence

8. The NSA or any other entity that has good, legal reason to do so will still be able to examine any part of this <u>far-flung</u> haystack. It simply will not hold the entire stack in a single server farm.

   A. remote   B. distant

   C. widely distributed   D. extensive

9. Without such safeguards, data can be <u>siphoned</u> off without anyone knowing.

   A. taken by sipping

   B. drawn off by a tube used to convey liquid from one container to another

   C. drawn in by a vacuum

   D. transferred over a period of time, especially illegally or unfairly

10. And in May the House of Representatives passed the U.S.A. Freedom Act; though considered weak by many privacy advocates, the bill would begin to restrict bulk data collection and introduce some <u>transparency</u> into the process.

    A. quality of permitting free passage of electromagnetic rays without distortion

    B. quality of allowing light to pass through so that objects behind can be distinctly seen

    C. being easy to perceive or detect

    D. quality of being open to public scrutiny

Unit 11
Cyber Security

- **Section Two**

Directions: There are two or three meanings for each semi-technical term underlined in the following sentences. Choose the correct one according to the context.

1. The NSA began collecting bulk phone call records from virtually every person in the U.S.; soon it was gathering data on bulk Internet traffic (vehicles moving on a public highway; the message or signals transmitted through a communication system; an action of dealing or trading in something illegal) from virtually everyone outside of the U.S.

2. Data hoarding (a stock or store of money or valued objects; an amassed store of useful information or facts, retained for future use) also made the NSA more vulnerable than ever to leaks.

3. Then NSA contractor Edward Snowden managed to download thousands of secret files from a server (a person or thing that provides a service or commodity; a computer or computer program which manages access to a centralized resource or service in a network; the player who serves in tennis and other racket sports) in Hawaii.

4. But a secretive agency collecting data on entire populations, storing those data in clandestine server farms (areas of land and its buildings used for growing crops and rearing animals; establishments at which something is produced or processed) and operating on them with little or no oversight is qualitatively different from anything that has come before.

5. For governments, it makes devastating leaks (a hole in a container through which contents, especially liquid or gas may accidently pass; an intentional disclosure of secret information) that much more likely.

6. There need be no rush to destroy the NSA's current data stores (shops of any size or kind; sheep, steers, cows or pigs acquired or kept for fattening; quantities or supplies of something kept for use as needed), because both the content of those records and the software associated with them will quickly become ancient history.

7. Each category of data-analysis operation or computing some statistic, whether it is searching for a particular item has its own characteristic pattern of communication—its own signature (a person's name written in a distinctive way as a form of identification in authorizing a check or document or concluding a letter; the action of signing a document; a distinctive pattern, product, or characteristic by which someone or

something can be identified) web of links and transmissions among databases.

8. All entities should be subject to metadata monitoring and investigative auditing (attending a class informally, without working for credit; conducting an official financial inspection of; conducting a systematic review of), similar to how credit cards are monitored for fraud today.

9. Several state governments in the U.S. are beginning to evaluate this architecture (the art or practice of designing and constructing buildings; the complex or carefully designed structure of something; the conceptual structure and logical organization of a computer or computer-based system) for both internal and external data-analysis services.

10. By maintaining a tamper-proof history of provenance and permissions, trust networks can be automatically audited to ensure that data-usage agreements are being honored (regard with great respect; fulfill an obligation; pay a check when due).

● **Section Three**

Directions: Match the Chinese terms with their English equivalents.

1. 身份凭证   A. big data
2. 联网移动设备   B. digital communication channel
3. 大数据   C. encryption
4. 网络犯罪   D. metadata
5. 连带责任   E. cybercrime
6. 物理数据库   F. data system
7. 黑客   G. identity credential
8. 数字时代   H. SWIFT
9. 数据分析   I. joint liability
10. 分布式数据   J. hacker
11. 数据系统   K. Internet-connected mobile device
12. 元数据   L. digital era
13. 加密   M. data analysis
14. 数字通信通道   N. distributing data
15. 环球银行金融电信协会   O. physical database

Unit 11
Cyber Security

■ III. Questions for Discussion

Directions: Work in groups and discuss the following questions.

1. Do you want to store your personal information in the cloud and share it with others? Why or why not?
2. As for cybersecurity, what should we do as individuals?
3. At the government's level, what can be done to ensure cyber security?

Part II Extensive Reading

Text B

How to Survive Cyberwar

Step one: Stop counting on others to protect you.

By Keren Elazari

Cybersecurity people like to say that there are two types of organizations—those that have been hit and those that do not know it yet. Recent headlines should prove that this joke is largely true. Cybercriminals stole the credit-card information and personal data of millions of people from companies that included Target, Home Depot and JPMorgan Chase. Security researchers discovered fundamental flaws in Internet building blocks, such as the so-called Heartbleed vulnerability in the popular OpenSSL cryptographic software library. A massive data-destruction attack sent Sony Pictures Entertainment back to using pen and paper. Criminals accessed the data of more than 80 million customers of health insurance giant Anthem. And these are just the incidents we know about.

In the coming years, cyberattacks will almost certainly intensify, and that is a problem for all of us. Now that everyone is connected in some way to cyberspace—through our phones, our laptops, our corporate networks—we are all vulnerable. Hacked networks, servers, personal computers and online accounts are a basic resource for cybercriminals and government snoops alike. Your corporate network or personal gaming PC can easily become another tool in the arsenal of criminals—or taxpayer-sponsored cyberspies. Compromised computers can be used as stepping-stones for the next attack or become part of a "botnet", a malicious network of controlled zombie devices rented out by the hour to launch denial-of-service attacks or distribute spam.

In response to threats such as these, the natural reflex of governments in the U.S. and elsewhere is to militarize cyberspace, to attempt to police the digital world using centralized bureaucracies and secret agencies. But this approach will never work. Cybersecurity is like a public health problem. Government agencies such as the Centers for Disease Control and Prevention have important roles to play, but they cannot stop the spread of diseases on their own. They can only do their job if citizens do theirs.

● The Vastness of Cyberspace

Part of the challenge of protecting cyberspace is that there is no single "cyberspace". To appreciate this fact, we must go back half a century, to the work of Norbert Wiener, a professor of mathematics at M.I.T. In 1948 Wiener borrowed from the ancient Greeks to describe a new scientific discipline he was developing: cybernetics, which he defined as the study of "control and communication in the animal and the machine". In the original Greek, *kybernetes* was the title for the steersman or the pilot directing and controlling naval vessels sailing in the Mediterranean. By analogy, cyberspace should be understood as the collection of interconnected electronic and digital technologies that enable control and communications of all systems underpinning modern life.

Cyberspace is not a public commons; it is not like international waters or the moon. It is not a collection of territories that governments or militaries could effectively control. Most of the technologies and networks that make up cyberspace are owned and maintained by multinational, for-profit conglomerates.

The number and variety of technologies included in this space are growing rapidly. Networking technology vendor Cisco Systems forecasts that by 2020, 50 billion devices will be connected to the Internet, including a large proportion of industrial-, military- and aerospace-related devices and systems. Each new thing that connects to cyber space is a potential target for a cyberattack, and attackers are good at finding the weakest links in any network. The hackers who breached Target's point-of-sale system and stole millions of payment cards, for example, gained access to the retailer's network by first hacking into an easier target: Fazio Mechanical Services, the refrigeration maintenance company that runs Target's heating and cooling systems.

The "things" of the Internet of Things are not just windows that attackers can sneak through: They are themselves targets for potential sabotage. As early as 2008, security researchers demonstrated that they could remotely hack into embedded pacemakers. Since then, hackers have shown that they can hijack implanted insulin pumps using radio signals, instructing the devices to dump insulin into patients' bloodstream, with potentially lethal results.

Physical infrastructure is also at risk of attack, as we learned in 2010, when the infamous computer virus Stuxnet was found to be responsible for widespread destruction of uranium-enrichment centrifuges inside a clandestine facility in Natanz, Iran. Stuxnet, allegedly the fruit of an intensive and costly collaboration between the U.S. and Israel, made a historic point: Digital computer code can disrupt and destroy analog, physical systems. Other attacks have since reinforced the point.

What this means is that cybersecurity is not just about securing computers, networks or Web servers. It is certainly not just about securing "secrets". The real battle in cyberspace is about protecting things, infrastructures and processes. The danger is the subversion and sabotage of the technologies we rely on every day. Cybersecurity is about protecting our way of life.

- **The Role of Government**

Governments face deep conflicts when it comes to securing cyberspace. Many federal agencies, including the Department of Homeland Security in the U.S., have an earnest interest in protecting national companies and citizens from cyberattacks. Yet other government entities can benefit from keeping the world's networks riddled with vulnerabilities.

One person's terrifying security vulnerability is another's secret weapon. Consider the Heartbleed bug. If you have used the Internet in the past five years, your information has probably been encrypted and decrypted by computers running OpenSSL software. SSL is the basic technology behind those "lock" icons we have grown to expect on secure Web sites. Heartbleed was the result of a basic software development error in one of OpenSSL's popular extensions, "Heartbeat", hence the name. When exploited, the bug gave eavesdroppers easy access to cryptographic keys, usernames and passwords, rendering moot any security offered by SSL encryption. OpenSSL was vulnerable for two years before two separate teams of security researchers discovered the bug. A few days later Bloomberg Businessweek cited anonymous sources claiming the NSA had been using the flaw to conduct cyberespionage for years.

Many of the world's leading powers have devoted their best tech talent and millions of dollars to finding and exploiting vulnerabilities such as Heartbleed. Governments also buy bugs on the open market, helping to sustain the trade in security flaws. A growing number of companies such as Vupen Security, a French firm, and Austin-based Exodus Intelligence specialize in the discovery and packaging of these precious bugs. In fact, some governments spend more money on researching and developing offensive cyber capabilities than they do on defensive cyber research. The Pentagon employs legions of vulnerability researchers, and the NSA reportedly spends two and a half times more money on offensive cyber research than on defense.

None of this is to say that governments are nefarious or that they are the enemies of cybersecurity. It is easy to see where agencies such as the NSA are coming from. Their job is to gather intelligence to prevent terrible acts; it makes sense that they would use any tool at their disposal to make that happen. Yet an important step in securing cyberspace is to honestly weigh the costs and benefits of government agencies cultivating vulnerabilities. Another key is to take full advantage of those things that governments can do and other organizations cannot. For example, they can enable or even compel companies and other organizations to share information about cyberattacks.

Banks in particular would benefit from sharing information about cyberattacks because attacks on financial institutions usually follow a predictable pattern: Once criminals find something that works on one bank, they try it on another bank and then another. Yet banks traditionally avoid disclosing information about attacks because it raises questions about their security. They also avoid talking to competitors; in some cases, antitrust laws prohibit them from doing so. Governments, however, can facilitate information sharing among banks. This is already happening in the U.S. in the form of the Financial Services Information Sharing and Analysis Center (FS-ISAC), which also serves global financial organizations.

● Hackers Can Help

As long as humans write code, vulnerabilities will exist. Driven by increasingly intense market pressures, technology companies push new products to market faster than ever before. In the past year, catalyzed by events such as the Edward Snowden NSA revelations, the technology industry and hacking community have become open to working together. Hundreds of companies now see the value of engaging hackers through so-called bug bounties and vulnerability reward programs, which offer incentives to independent researchers who report vulnerabilities and security problems. Netscape Communications created the first bug bounty program in 1995 as a way to find flaws in the Netscape Navigator Web browser. Today, 20 years later, research has shown that the strategy is one of the more cost-effective measures the organization and its successor, Mozilla, have taken to bolster security. Private and public communities of security professionals share information about malware, threats and vulnerabilities to create a kind of distributed immune system.

As cyberspace expands, car manufacturers, medical device companies, home-entertainment-system providers and other businesses will have to start thinking like cybersecurity firms. Here, too, the hacker community can help. In 2013, for example, security experts Joshua Corman and Nicholas Percoco launched a movement called "I Am the Cavalry", urging hackers to conduct responsible security research that makes a difference in the world, with an emphasis on critical areas such as public infrastructures and automotive, medical device and connected home technologies.

The good news is that this distributed immune system is growing stronger. In January, Google launched a new program that complements its bug bounty program, offering grants to encourage security researchers to scrutinize the company's products. The program is an admission that even companies with the best in-house tech talent on the planet could use the outside perspective of friendly hackers.

The bad news is that some elements of the cybersecurity approach the Obama administration is pursuing could effectively criminalize common vulnerability research practices and tools, weakening this developing immune system. Many in the security community fear that both the current version of the Computer Fraud and Abuse Act and proposed changes to the law define hacking so expansively that even clicking on a link to a Web site containing leaked or stolen information could be considered trafficking in stolen goods. Criminalizing the work of independent security researchers would harm us all and have little effect on criminals motivated by profit or ideology.

- **Individual Responsibility**

The next few years could be messy. We will see more data breaches, and we will almost certainly see a vigorous debate about how much control over the digital realm we should cede to governments in return for security. The truth is that securing cyberspace will require solutions from many realms: technical, legal, economic and political. It is also up to us, the general public. As consumers, we should demand that companies make their products more secure. As citizens, we should hold our governments accountable when they intentionally weaken security. And as individual points of potential failure, we have a responsibility to secure our own stuff.

Defending ourselves involves simple steps such as keeping our software up-to-date, using secure Web browsers, and enabling two-factor authentication on our e-mail and social-media accounts. But it also involves being aware that each of our devices is a node in a much larger system and that the little choices we make can have wide-ranging effects.

(Excerpt from *Scientific American*, April 2015)

Exercises

I. Translate the following technical terms into English.

1. 网络攻击 _____
2. 埋入式起搏器 _____
3. 拒绝服务 _____
4. 垃圾邮件 _____
5. 僵尸网络 _____
6. 控制论 _____
7. 胰岛素泵 _____
8. 铀浓缩离心机 _____
9. 销售点系统 _____
10. 密钥 _____
11. 加密套接字协议层 _____
12. 解密 _____

13. 网络侦察 _____   14. 图标 _____
15. 网络安全_____   16. 社交网络账号 _____
17. 漏洞报告奖励 _____ 18. 分布式免疫系统 _____
19. 恶意软件 _____   20. 双因素身份认证 _____

II. Translate the following paragraphs into Chinese.

The next few years could be messy. We will see more data breaches, and we will almost certainly see a vigorous debate about how much control over the digital realm we should cede to governments in return for security. The truth is that securing cyberspace will require solutions from many realms: technical, legal, economic and political. It is also up to us, the general public. As consumers, we should demand that companies make their products more secure. As citizens, we should hold our governments accountable when they intentionally weaken security. And as individual points of potential failure, we have a responsibility to secure our own stuff.

Defending ourselves involves simple steps such as keeping our software up-to-date, using secure Web browsers, and enabling two-factor authentication on our e-mail and social-media accounts. But it also involves being aware that each of our devices is a node in a much larger system and that the little choices we make can have wide-ranging effects. Again, cybersecurity is just like public health. Wash your hands and get vaccinated, and you can avoid spreading the disease further.

Text C

In Tech We Don't Trust

Tech companies promise the world, but how do we know that we're not the ones being sold out?

By David Pogue

Last October, T-Mobile made an astonishing announcement: From now on, when you travel internationally with a T-Mobile phone, you get free unlimited text messages and Internet use. Phone calls to any country are 20 cents a minute.

T-Mobile's plan changes everything. It ends the age of putting your phone in airplane mode overseas, terrified by tales of $6,000 overage charges. I figured my readers would be jubilant. But a surprising number had a very different reaction. "Why should I believe them?" they wrote. "Cell carriers have lied to us for years."

That's not the first time that promises from a tech company have been greeted not with joy but with skepticism. When Apple introduced a fingerprint scanner into the Home button of the iPhone 5S, you might have expected the public's reaction to be, "Wow, that's much faster than having to type in a password 50 times a day!" But instead a common reaction was: "Oh, great. So now Apple can give my fingerprints to the NSA."

Really? That's your reaction to the first cell phone with a finger scanner that actually works?

And it's not so unreasonable. Technology used to be admired in America. We marveled at the first radio, the laptop computer, the flat TV. Tech companies were our blue-chip companies. An IBM man was a good catch—respected, impressive. We were proud of our technological prowess and of the companies that were at the forefront.

Today it's not so simple. Our tech companies have a trust problem.

Over the years they've brought it on themselves. Google tested privacy tolerance when it introduced Gmail—with ads relating to the content of your messages. (It doesn't seem to matter that software algorithms, not people, scan your mail.)

Then a team of researchers discovered that when you synced your iPhone, your computer downloaded a log of your geographical movements, in a form accessible with simple commands. (Apple quickly revised its software.) When Barnes & Noble understated the weight of its Nook e-reader in 2010 or overstated the resolution of the Nook in 2011, suddenly even product specs could no longer be trusted.

Next came news about the National Security Agency and its collection of e-mail correspondence, chat transcripts and other data from Microsoft, Google, Facebook, Apple and others. Those companies admit to complying with the occasional warrant for individuals' data, but they strenuously deny providing the NSA with bigger sets of data. Do you think that makes the news any easier to take?

Of course not. We're human. We look for patterns. Each new headline further shakes our trust in the whole system.

These days tech companies make efforts to respect, or at least to humor, the public's alarm. In the latest iPhone software, for example, Apple has provided an almost hilariously complete set of on/off switches, one for every app that might want access to your location information.

But it may be too late for that. These companies' products are impossibly complex. There's no way for an individual to verify that software does exactly what we think it does. How do we know those iOS 7 switches do anything at all?

Every time a company slips up, we can only assume that it is just the tip of the iceberg. It may take years for these companies to regain our trust.

But this "I don't trust them anymore" thing sounds distinctly familiar. And it isn't specific to tech companies. At one time or another, haven't we also learned not to trust our government? Our police? Our hospitals? Our newspapers? Our medicines? And, goodness knows, our phone companies?

It's too bad. Mistrust means a life of wariness. It means constant psychic energy, insecurity, less happiness. And then, when we finally get what should be terrific news from a tech company, we're deprived of that little burst of unalloyed pleasure.

(Excerpt from *Scientific American*, January 2014)

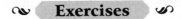

Exercises

Reading Comprehension

Directions: Answer the following questions based on the information from the text.

1. What were people's attitudes towards the T-Mobile announcement?
2. When Apple introduced a fingerprint scanner into the Home button of iPhone 5S, what was the common reaction and why?
3. What has caused a trust problem for tech companies?
4. Do you like to be accessed to your location information? Why or why not?
5. What does "unalloyed pleasure" mean?

Unit 12
Science and Society

> **导读**
>
> 本单元的文章均涉及技术与社会之间的种种联系。Text A从进化的角度详细讲解了未来政府走向透明化的趋势。信息透明让社会进化，无法适应新形势的游戏者很快就会遭到淘汰。Text B则讲述了信息时代背景下，隐私与人类行为之间的关系。Text C详细阐述了现代数字社会中被遗忘权的重要意义。

Part I Intensive Reading

Text A

Our Transparent Future

No secret is safe in the digital age. The implications for our institutions are downright Darwinian.

By Daniel C. Dennett and Deb Roy

More than half a billion years ago a spectacularly creative burst of biological innovation called the Cambrian explosion❶ occurred. In a geologic "instant" of several million years, organisms developed strikingly new body shapes, new organs, and new predation strategies and defenses against them. Evolutionary biologists disagree about what triggered this prodigious wave of novelty, but a particularly

---

❶ The Cambrian explosion was the relatively short evolutionary event, beginning around 541 million years ago in the Cambrian period, during which most major animal phyla appeared, as indicated by the fossil record. Additionally, the event was accompanied by major diversification of other organisms. Prior to the Cambrian explosion, most organisms were simple, composed of individual cells occasionally organized into colonies. Over the following 70 to 80 million years, the rate of diversification accelerated by an order of magnitude and the diversity of life began to resemble that of today.

compelling hypothesis, advanced by University of Oxford zoologist Andrew Parker, is that light was the trigger. Parker proposes that around 543 million years ago, the chemistry of the shallow oceans and the atmosphere suddenly changed to become much more transparent. At the time, all animal life was confined to the oceans, and as soon as the daylight flooded in, eyesight became the best trick in the sea. As eyes rapidly evolved, so did the behaviors and equipment that responded to them.

Whereas before all perception was proximal—by contact or by sensed differences in chemical concentration or pressure waves—now animals could identify and track things at a distance. Predators could home in on their prey; prey could see the predators coming and take evasive action. Locomotion is a slow and stupid business until you have eyes to guide you, and eyes are useless if you cannot engage in locomotion, so perception and action evolved together in an arms race. This arms race drove much of the basic diversification of the tree of life we have today.

Parker's hypothesis about the Cambrian explosion provides an excellent parallel for understanding a new, seemingly unrelated phenomenon: the spread of digital technology. Although advances in communications technology have transformed our world many times in the past, digital technology could have a greater impact than anything that has come before. It will enhance the powers of some individuals and organizations while subverting the powers of others, creating both opportunities and risks that could scarcely have been imagined a generation ago.

Through social media, the Internet has put global-scale communications tools in the hands of individuals. A wild new frontier has burst open. Services such as YouTube, Facebook, Twitter, Tumblr, Instagram, WhatsApp and SnapChat generate new media on a par with the telephone or television—and the speed with which these media are emerging is truly disruptive. It took decades for engineers to develop and deploy telephone and television networks, so organizations had some time to adapt. Today a social-media service can be developed in weeks, and hundreds of millions of people can be using it within months. This intense pace of innovation gives organizations no time to adapt to one medium before the arrival of the next.

The tremendous change in our world triggered by this media inundation can be summed up in a word: transparency. The age-old game of hide-and-seek that has

Unit 12
Science and Society

shaped all life on the planet has suddenly shifted its playing field, its equipment and its rules. The players who cannot adjust will not last long.

The impact on our organizations and institutions will be profound. Governments, armies, churches, universities, banks and companies all evolved to thrive in a relatively murky epistemological environment, in which most knowledge was local, secrets were easily kept, and individuals were, if not blind, myopic. When these organizations suddenly find themselves exposed to daylight, they quickly discover that they can no longer rely on old methods; they must respond to the new transparency or go extinct. Just as a living cell needs an effective membrane to protect its internal machinery from the vicissitudes of the outside world, so human organizations need a protective interface between their internal affairs and the public world, and the old interfaces are losing their effectiveness.

• Claws, Jaws and Shells

In his 2003 book, *In the Blink of an Eye*, Parker argues that the external, hard body parts of fauna responded most directly to the riot of selection pressures of the Cambrian explosion. The sudden transparency of the seas led to the emergence of camera-style retinas, which in turn drove rapid adaptation of claws, jaws, shells and defensive body parts. Nervous systems evolved, too, as animals developed new predatory behaviors and, in response, methods of evasion and camouflage.

By analogy, we might expect organizations to respond to the pressure of digitally driven social transparency with adaptations in their external body parts. In addition to the organs they use to deliver goods and services, these body parts include information-handling organs of control and self-presentation: public relations, marketing and legal departments, for instance. It is here we can see the impact of transparency most directly. Organizations that need weeks or months to develop communications strategies gated by slow-moving legal departments will find themselves quickly out of sync. Old habits must be rewired, or else the organization will fail.

Easier access to data has enabled new forms of public commentary grounded in comprehensive empirical observations. Data journalist Nate Silver demonstrated as much during the 2012 U.S. presidential elections. While some news organizations spun why our candidate will win narratives based on cherry-picked polling data, Silver gave us explanatory narratives grounded in all polling data. Not only did Silver predict the elections with uncanny accuracy, but by openly sharing his methodology, he also eliminated any doubt that he merely got lucky. With transparent public polls increasingly available, news organizations and political analysts that spin selectively grounded stories are going to face an increasingly difficult existence.

Consumer goods manufacturers face a closely related challenge. User reviews of products and services are changing the balance of power between customers and companies. Small groups of people with shared values, beliefs and goals—particularly those who can coordinate quickly in a crisis using ad hoc❶ channels of internal communication—will be best at the kind of fast, open, responsive communication the new transparency demands. As the pressures of mutual transparency increase, we will either witness the evolution of novel organizational arrangements that are far more decentralized than today's large organizations, or we will find that Darwinian pressures select for smaller organizations, heralding an era of "too big to succeed".

● **The Half-Lives of Secrets**

U.S. supreme court justice Louis D. Brandeis, an early champion of transparency, is often quoted on the topic. "Sunlight is said to be the best of disinfectants," he famously wrote. He was right, of course, both metaphorically and literally. But sunlight can be dangerous, too. What if in our zeal for purification, we kill too many friendly cells? What about the risk of destroying the integrity or effectiveness of organizations by exposing too much of their inner workings to the world?

Brandeis was an enemy of secrecy. He apparently thought that the more transparent institutions became, the better they would be. A biological perspective helps us see that transparency is a mixed blessing. Animals, even plants, can be seen to be agents with agendas. Informed by their sensory organs, these agents act to further their own welfare. A human organization is similar. It is an agent composed

---

❶ Ad hoc is a Latin phrase meaning "for this". In English, it generally signifies a solution designed for a specific problem or task, non-generalizable, and not intended to be adapted to other purposes.

Unit 12
Science and Society

of large numbers of working, living parts—people. But unlike the cells that make up plants and animals, people have wide interests and perceptual abilities.

It was not always so. In earlier times, dictators could rule quite inscrutably from behind high walls, relying on hierarchical organizations composed of functionaries with very limited knowledge of the organization of which they were a part and even less information about the state of the world, near and far. Churches have been particularly adept at thwarting the curiosity of their members, keeping them uninformed or misinformed about the rest of the world while maintaining a fog of mystery around their internal operations, histories, finances and goals. Armies have always benefited from keeping their strategies secret—not just from the enemy but from the troops as well. Soldiers who learn the anticipated casualty rates of an operation will not be as effective as those who remain oblivious about their likely fate. Moreover, if an uninformed soldier is captured, he will have less valuable information to divulge under interrogation.

One of the fundamental insights of game theory[1] is that agents must keep secrets. An agent who reveals "state" to another agent has lost some valuable autonomy and is in danger of being manipulated. To compete fairly in an open market, manufacturers need to protect the recipes for their products, their expansion plans and other proprietary information. Schools and universities need to keep their examinations secret until the students take them. President Barack Obama promised a new era of government transparency, but despite significant improvements, large arenas of secrecy and executive privilege are enforced as vigorously as ever. This is as it should be. Economic statistics, for instance, need to be kept secret until they are officially revealed to prevent insider exploitation. A government needs a poker face[2] to conduct its activities, but the new transparency makes this harder than ever before.

Edward Snowden's revelations about the inner workings of the National Security

---

[1] Game theory is the study of mathematical models of conflict and cooperation between intelligent rational decision-makers. Game theory is mainly used in economics, political science, and psychology, as well as logic, computer science, biology and poker.

[2] Poker face is a deliberately-induced blank expression meant to conceal one's emotions, a common practice of maintaining one's composure when playing the card game poker.

Agency demonstrate how in the era of transparency, a single whistle-blower❶ or mole can disrupt a massive organization. Although Snowden used traditional news organizations to leak information, social-media reaction and amplification assured that the news stories would not die, putting sustained widespread pressure on the NSA and the federal government to act.

The NSA's outer "skin" is adapting dramatically in response. The mere fact that the agency publicly defended itself against Snowden's accusations was unprecedented for an organization that has long resided behind a veil of complete secrecy. As Joel Brenner, former senior counsel at the NSA, reflected on the sudden shift of the agency's operating environment at a December 2013 panel hosted at the M.I.T. Media Lab, "Very few things will be secret anymore, and those things which are kept secret won't stay secret very long…The real goal in security now is to retard the degradation of the half-lives of secrets. Secrets are like isotopes."

● Information Chaff

In their evolutionary arms race, the Cambrian fauna invented a bounty of evasive measures and countermeasures, and this arsenal of tricks has grown ever since. Animals have developed camouflage, alarm calls to warn of approaching threats, bright markings that falsely advertise them to potential predators as being poisonous. The new transparency will lead to a similar proliferation of tools and techniques for information warfare: campaigns to discredit sources, preemptive strikes, stings, and more.

Nature has inspired devious armaments before. The cloud of ink released by cephalopods fleeing a predator was reinvented in aerial warfare as chaff—confusing clouds of radar-reflective metal scraps or dummy warheads that could attract defensive missiles. We can predict the introduction of chaff made of nothing but megabytes of misinformation. It will quickly be penetrated, in turn, by more sophisticated search engines, provoking the generation of ever more convincing chaff. Encryption and decryption schemes will continue to proliferate as well, as

---

❶ Whistle-blower refers to a person who exposes any kind of information or activity that is deemed illegal, unethical, or not correct within an organization that is either private or public.

organizations and individuals struggle to preserve their privacy and reputations.

• Speciation of Organizations

A final implication of our Cambrian analogy is that we should soon witness a massive diversification of species of organizations. In the U.S., a new class of corporation, the B Corp, was recently created to recognize the need for ventures with double bottom lines optimized for both profit and social purpose. Google and Facebook broke with tradition by enacting unusually powerful voting rights for their founders, yielding publicly traded companies that remain privately controlled, enabling the founders to steer their companies based on their longterm plans with relative indifference to the quarterly whims of Wall Street. The organized protests during the Arab Spring❶, enabled by social media and unrivaled in their combination of scale and speed of formation, are perhaps also a new kind of (ephemeral) human organization. Time will tell, but it appears that we might be at the cusp of a radical branching of the organizational tree of life.

The speed with which transparency will shape an organization depends on its competitive niche. Commercial companies are most exposed to the effects of public opinion because customers can easily switch to alternatives.

Most sheltered from immediate evolutionary pressures are systems of government. Protests fueled by social media can topple rulers and ruling parties, but the underlying organs of the state tend to continue relatively unperturbed by changes in political leadership. State machinery faces little competitive pressure and is thus the slowest to evolve. Under popular pressure, governments are opening access to vast new streams of raw data produced by their internal operations. Coupled with advances in largescale pattern analysis, data visualization, and datagrounded professional and citizen journalism, we are creating powerful social feedback loops that will accelerate transparency of organizations.

There is a selflimiting aspect to this emerging new human order. Just as ant

---

❶ Arab Spring, or the Democracy Spring, refers to a revolutionary wave of both violent and non-violent demonstrations, protests, riots, coups and civil wars in the Arab world that began on 17 December 2010 in Tunisia with the Tunisian Revolution, and spread throughout the countries of the Arab League and its surroundings.

colonies can do things that individual ants cannot, human organizations can also transcend the abilities of individuals, giving rise to superhuman memories, beliefs, plans, actions—perhaps even superhuman values. For better or for worse, however, we are on an evolutionary course to rein in our superhuman organizations by holding them accountable to individual human standards.

(Excerpt from Scientific American, March 2015)

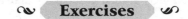

Exercises

I. Reading Comprehension

● **Section One**

Directions: Answer the following questions based on the information from the text.

1. What is the Cambrian analogy mentioned in the text?
2. What are the major implications of the Cambrian analogy?
3. What is your understanding of the statement that "transparency is a mixed blessing"?
4. What is the fundamental insight of game theory?
5. What impact does the digital technology have on the modern society?

● **Section Two**

Directions: Write an abstract based on the text in no more than 200 words.

Abstract:
Key words:

Unit 12
Science and Society

II. Vocabulary

• **Section One**

Directions: Choose the explanation that is closest in meaning to the underlined part in each sentence.

1. At the time, all animal life was confined to the oceans, and as soon as the daylight flooded in, eyesight became the best trick in the sea.
    A. something you do in order to deceive someone
    B. something you do to surprise someone and to make other people laugh
    C. a skillful set of actions that seem like magic, done to entertain people
    D. a way of doing something that works very well but may not be easy to notice
2. Locomotion is a slow and stupid business until you have eyes to guide you, and eyes are useless if you cannot engage in locomotion, so perception and action evolved together in an arms race.
    A. a powered railway vehicle used for pulling trains
    B. movement or the ability to move from one place to another
    C. a word or phrase, especially with regard to style or idiom
    D. a large and mainly tropical grasshopper with strong powers of flight
3. Parker's hypothesis about the Cambrian explosion provides an excellent parallel for understanding a new, seemingly unrelated phenomenon: the spread of digital technology.
    A. a person or thing that is similar or analogous to another
    B. a corresponding line on a map
    C. side by side and having the same distance continuously between them
    D. computing involving the simultaneous performance of operations
4. Just as a living cell needs an effective membrane to protect its internal machinery from the vicissitudes of the outside world, so human organizations need a protective interface between their internal affairs and the public world, and the old interfaces are losing their effectiveness.
    A. a change of circumstances or fortune
    B. an act of defeating an enemy or opponent in a battle, game, or other competition
    C. the area near or surrounding a particular place

D. a person harmed, injured, or killed as a result of a crime, accident, or other event or action

5. While some news organizations spun whyourcandidatewillwin narratives based on cherrypicked polling data, Silver gave us explanatory narratives grounded in all polling data.

   A. select the ripest and healthiest fruits

   B. selectively choose from what is available

   C. hold something dear

   D. keep in one's mind (a hope or ambition)

6. U.S. supreme court justice Louis D. Brandeis, an early champion of transparency, is often quoted on the topic.

   A. a person who has defeated all rivals in a competition, especially a sporting contest

   B. a person who fights or argues for a cause or on behalf of another

   C. someone who competes with or fights another in a contest, game, or argument

   D. a spectator

7. Edward Snowden's revelations about the inner workings of the National Security Agency demonstrate how in the era of transparency, a single whistleblower or mole can disrupt a massive organization.

   A. a small burrowing insectivorous mammal with dark velvety fur, a long muzzle, and very small eyes

   B. someone within an organization who anonymously betrays confidential information

   C. a small, often slightly raised blemish on the skin made dark by a high concentration of melanin

   D. a large solid structure on a shore serving as a pier, breakwater, or causeway

8. The speed with which transparency will shape an organization depends on its competitive niche.

   A. a shallow recess, especially one in a wall to display a statue or other ornament

   B. a fine detail or distinction

   C. a daughter of one's brother or sister

   D. a position or role taken by a kind of organism within its community

9. To compete fairly in an open market, manufacturers need to protect the recipes for their products, their expansion plans and other proprietary information.

Unit 12
Science and Society

A. a person or thing that receives or is awarded something

B. a medical prescription

C. something which is likely to lead to a particular outcome

D. a set of instructions for preparing a particular dish

10. State machinery faces little competitive pressure and is thus the slowest to evolve.

    A. machines collectively

    B. the components of a machine

    C. the organization or structure of something or for doing something

    D. a well-organized group of powerful people

● **Section Two**

Directions: There are two or three meanings for each semi-technical term underlined in the following sentences. Choose the correct one according to the context.

1. The players (people taking part in a sport or game; people who are involved and influential in an area or activity) who cannot adjust will not last long.
2. Just as a living cell needs an effective membrane to protect its internal machinery from the vicissitudes of the outside world, so human organizations need a protective interface (a point where two systems, subjects, or organizations meet and interact; a surface forming a common boundary between two objects or liquids) between their internal affairs and the public world.
3. In his 2003 book, *In the Blink of an Eye*, Parker argues that the external, hard body parts of fauna responded most directly to the riot of selection (a number of carefully chosen things; a process in which environmental or genetic influences determine which types of organism thrive better than others, regarded as a factor in evolution) pressures of the Cambrian explosion.
4. Nervous systems evolved, too, as animals developed new predatory behaviors and, in response, methods of evasion and camouflage (the disguising of military personnel, equipment, and installations by painting or covering them to make them blend in with their surroundings; the natural coloring that enables it to blend in with its surroundings; actions or devices intended to disguise or mislead).
5. Soldiers who learn the anticipated casualty rates of an operation (an act of surgery

performed on a patient; a piece of organized and concerted activity involving members of the armed forces or the police; the fact or condition of functioning or being active) will not be as effective as those who remain oblivious about their likely fate.

6. Parker proposes that around 543 million years ago, the chemistry (scientific study of substances and of the way they react with other substances; chemical composition and properties of a substance; a phenomenon perceived as complex or mysterious) of the shallow oceans and the atmosphere suddenly changed to become much more transparent.

7. Although Snowden used traditional news organizations to leak information, social-media reaction and amplification (increase of the sound volume; the quality of becoming more marked or intense; increase of amplitude of an electrical signal) assured that the news stories would not die.

8. As Joel Brenner reflected on the sudden shift of the agency's operating environment at a December 2013 panel (a flat board on which instruments or controls are fixed; a small group of people brought together to discuss, investigate, or decide upon a particular matter; a list of available jurors or a jury) hosted at the M.I.T. Media Lab, "Very few things will be secret anymore, and those things which are kept secret won't stay secret very long."

9. The real goal in security now is to retard the degradation (reduction of energy to a less readily convertible form; wearing down and disintegrating of rock; lowering the quality of something) of the halflives of secrets. Secrets are like isotopes.

10. Just as ant colonies (countries or areas under the full or partial political control of another country; communities of animals of one kind living close together or forming a physically connected structure) can do things that individual ants cannot, human organizations can also transcend the abilities of individuals.

Unit 12
Science and Society

● **Section Three**

Directions: Match the Chinese terms with their English equivalents.

1. 膜   A. methodology
2. 视网膜   B. chemical concentration
3. 消毒剂   C. sensory organ
4. 动物群   D. game theory
5. 认知论的   E. Snapchat
6. 化学浓度   F. retina
7. 气压波   G. fauna
8. 经验观测值   H. dummy warhead
9. 内部工作原理   I. arms race
10. 阅后即焚   J. epistemological
11. 方法论   K. membrane
12. 假弹头   L. empirical observation
13. 博弈论   M. pressure wave
14. 军备竞赛   N. inner working
15. 感觉器官   O. disinfectant

III. Questions for Discussion

Directions: Work in groups and discuss the following questions

1. What is the good and the bad for a transparent society?
2. What is the gap between the advocacy of transparency and the reality in Western countries?
3. What types of organizations do you think the new transparency will ultimately weed out?

Part II Extensive Reading

Text B

Privacy and Human Behavior in the Age of Information

By Alessandro Acquisti, Laura Brandimarte, and George Loewenstein

This Review summarizes and draws connections between diverse streams of empirical research on privacy behavior. We use three themes to connect insights from social and behavioral sciences: people's uncertainty about the consequences of privacy-related behaviors and their own preferences over those consequences; the context-dependence of people's concern, or lack thereof, about privacy; and the degree to which privacy concerns are malleable—manipulable by commercial and governmental interests. Organizing our discussion by these themes, we offer observations concerning the role of public policy in the protection of privacy in the information age.

If this is the age of information, then privacy is the issue of our times. Activities that were once private or shared with the few now leave trails of data that expose our interests, traits, beliefs, and intentions. We communicate using e-mails, texts, and social media; find partners on dating sites; learn via online courses; seek responses to mundane and sensitive questions using search engines; read news and books in the cloud; navigate streets with geotracking systems; and celebrate our newborns, and mourn our dead, on social media profiles. Through these and other activities, we reveal information—both knowingly and unwittingly—to one another, to commercial entities, and to our governments. The monitoring of personal information is ubiquitous; its storage is so durable as to render one's past undeletable —a modern digital skeleton in the closet❶. Accompanying the acceleration in data collection are steady advancements in the ability to aggregate, analyze, and draw sensitive

---

❶ Skeleton in the closet, or skeleton in the cupboard, is a colloquial phrase and idiom used to describe an undisclosed fact about someone which, if revealed, would have a negative impact on perceptions of the person, such as having a corpse concealed in your home long enough for it to decompose into bones. "Cupboard" is used in British English instead of the American English word "closet".

inferences from individuals' data.

Both firms and individuals can benefit from the sharing of once hidden data and from the application of increasingly sophisticated analytics to larger and more interconnected databases. The erosion of privacy can threaten our autonomy, not merely as consumers but as citizens.

Because of the seismic nature of these developments, there has been considerable debate about individuals' ability to navigate a rapidly evolving privacy landscape, and about what should be done about privacy at a policy level. Traditional tools for privacy decision-making such as choice and consent no longer provide adequate protection. Instead of individual responsibility, regulatory intervention may be needed to balance the interests of the subjects of data against the power of commercial entities and governments holding that data.

Are individuals up to the challenge of navigating privacy in the information age? To address this question, we review diverse streams of empirical privacy research from the social and behavioral sciences. We highlight factors that influence decisions to protect or surrender privacy and how, in turn, privacy protections or violations affect people's behavior. Information technologies have progressively encroached on every aspect of our personal and professional lives. Thus, the problem of control over personal data has become inextricably linked to problems of personal choice, autonomy, and socioeconomic power. Accordingly, this Review focuses on the concept of, and literature around, informational privacy, but also touches on other conceptions of privacy, such as anonymity or seclusion.

We use three themes (uncertainty, context-dependence, and malleability) to organize and draw connections between streams of privacy research that, in many cases, have unfolded independently. These three themes are closely connected. Context-dependence is amplified by uncertainty. Because people are often "at sea" when it comes to the consequences of, and their feelings about, privacy, they cast around for cues to guide their behavior. Privacy preferences and behaviors are, in turn, malleable and subject to influence in large part because they are context-dependent and because those with an interest in information divulgence are able to manipulate context to their advantage.

● **Uncertainty**

Individuals manage the boundaries between their private and public spheres in numerous ways: via separateness, reserve, or anonymity; by protecting personal information; but also through deception and dissimulation. People establish such boundaries for many reasons, including the need for intimacy and psychological respite and the desire for protection from social influence and control. However, at other times, people experience considerable uncertainty about whether, and to what degree, they should be concerned about privacy.

A first and most obvious source of privacy uncertainty arises from incomplete and asymmetric information. Advancements in information technology have made the collection and usage of personal data often invisible. As a result, individuals rarely have clear knowledge of what information other people, firms, and governments have about them or how that information is used and with what consequences. To the extent that people lack such information, or are aware of their ignorance, they are likely to be uncertain about how much information to share.

Two factors exacerbate the difficulty of ascertaining the potential consequences of privacy behavior. First, whereas some privacy harms are tangible, such as the financial costs associated with identity theft, many others, such as having strangers become aware of one's life history, are intangible. Second, privacy is rarely an unalloyed good; it typically involves trade-offs. For example, ensuring the privacy of a consumer's purchases may protect her from price discrimination but also deny her the potential benefits of targeted offers and advertisements.

A second source of privacy uncertainty relates to preferences. Even when aware of the consequences of privacy decisions, people are still likely to be uncertain about their own privacy preferences. Research on preference uncertainty shows that individuals often have little sense of how much they like goods, services, or other people.

The remarkable uncertainty of privacy preferences comes into play in efforts to measure individual and group differences in preference for privacy. For example, Westin famously used broad (that is, not contextually specific) privacy questions in surveys to cluster individuals into privacy segments: privacy fundamentalists, pragmatists, and unconcerned. When asked directly, many people fall in the first segment: They profess to care a lot about privacy and express particular concern over losing control of their personal information or others gaining unauthorized access to it. This discrepancy between attitudes and behaviors has become known as the "privacy paradox".

In one early study illustrating the paradox, participants were first classified into categories of privacy concern inspired by Westin's categorization based on their responses to a survey dealing with attitudes toward sharing data. Next, they were presented with products to purchase at a discount with the assistance of an anthropomorphic shopping agent. Few, regardless of the group they were categorized in, exhibited much reluctance to answering the increasingly sensitive questions the agent plied them with.

Why do people who claim to care about privacy often show little concern about it in their daily behavior? One possibility is that the paradox is illusory. Thus, one might care deeply about privacy in general but, depending on the costs and benefits prevailing in a specific situation, seek or not seek privacy protection.

This explanation for the privacy paradox, however, is not entirely satisfactory for two reasons. The first is that it fails to account for situations in which attitude-behavior dichotomies arise under high correspondence between expressed concerns and behavioral actions. For example, one study compared attitudinal survey answers to actual social media behavior. Even within the subset of participants who expressed the highest degree of concern over strangers being able to easily find out their sexual orientation, political views, and partners' names, 48% did in fact publicly reveal their sexual orientation online, 47% revealed their political orientation, and 21% revealed their current partner's name. The second reason is that privacy decision-making is only in part the result of a rational "calculus" of costs and benefits; it is also affected by misperceptions of those costs and benefits, as well as social norms, emotions, and heuristics. Any of these factors may affect behavior differently from how they affect attitudes.

Preference uncertainty is evident not only in studies that compare stated attitudes with behaviors, but also in those that estimate monetary valuations of privacy.

"Explicit" investigations ask people to make direct trade-offs, typically between privacy of data and money. For instance, in a study conducted both in Singapore and the United States, students made a series of hypothetical choices about sharing information with Web sites that differed in protection of personal information and prices for accessing services. Using conjoint analysis, the authors concluded that subjects valued protection against errors, improper access, and secondary use of personal information between $30.49 and $44.62. Similar to direct questions about attitudes and intentions, such explicit investigations of privacy valuation spotlight privacy as an issue that respondents should take account of and, as a result, increase the weight they place on privacy in their responses.

Implicit investigations, in contrast, infer valuations of privacy from day-to-day decisions in which privacy is only one of many considerations and is typically not highlighted. Individuals engage in privacy-related transactions all the time, even when the privacy trade-offs may be intangible or when the exchange of personal data may not be a visible or primary component of a transaction.

In fact, attempts to pinpoint exact valuations that people assign to privacy may be misguided, as suggested by research calling into question the stability, and hence validity, of privacy estimates. In one field experiment inspired by the literature on endowment effects, shoppers at a mall were offered gift cards for participating in a non-sensitive survey. The cards could be used online or in stores, just like debit cards. Participants were given either a $10 "anonymous" gift card (transactions done with that card would not be traceable to the subject) or a $12 trackable card (transactions done with that card would be linked to the name of the subject). Initially, half of the participants were given one type of card, and half the other. Then, they were all offered the opportunity to switch. Some shoppers, for example, were given the anonymous $10 card and were asked whether they would accept $2 to "allow my name to be linked to transactions done with the card"; other subjects were asked whether they would accept a card with $2 less value to "prevent my name from being linked to transactions done with the card". Of the subjects who originally held the less valuable but anonymous card, five times as many (52.1%) chose it and kept it over the other card than did those who originally held the more valuable card (9.7%).

This suggests that people value privacy more when they have it than when they do not.

The consistency of preferences for privacy is also complicated by the existence of a powerful countervailing motivation: the desire to be public, share, and disclose. Social penetration theory suggests that progressively increasing levels of self-disclosure are an essential feature of the natural and desirable evolution of interpersonal relationships from superficial to intimate. Such a progression is only possible when people begin social interactions with a baseline level of privacy. Paradoxically, therefore, privacy provides an essential foundation for intimate disclosure. Similar to privacy, self-disclosure confers numerous objective and subjective benefits, including psychological and physical health. The desire for interaction, socialization, disclosure, and recognition or fame are human motives no less fundamental than the need for privacy. The electronic media of the current age provide unprecedented opportunities for acting on them. Through social media, disclosures can build social capital, increase self-esteem, and fulfill ego needs.

- **Context-Dependence**

Much evidence suggests that privacy is a universal human need. However, when people are uncertain about their preferences they often search for cues in their environment to provide guidance. And because cues are a function of context, behavior is as well. Applied to privacy, context-dependence means that individuals can, depending on the situation, exhibit anything ranging from extreme concern to apathy about privacy. Adopting the terminology of Westin, we are all privacy pragmatists, privacy fundamentalists, or privacy unconcerned, depending on time and place.

The way we construe and negotiate public and private spheres is context-dependent because the boundaries between the two are murky: The rules people follow for managing privacy vary by situation, are learned over time, and are based on cultural, motivational, and purely situational criteria. The theory of contextual "integrity" posits that social expectations affect our beliefs regarding what is private and what is public, and that such expectations vary with specific contexts. Thus, seeking privacy in public is not a contradiction; individuals can manage privacy even while sharing information, and even on social media.

The cues that people use to judge the importance of privacy sometimes result in sensible behavior. For instance, the presence of government regulation has been shown to reduce consumer concern and increase trust; it is a cue that people use to infer the existence of some degree of privacy protection. In other situations, however, cues can be unrelated, or even negatively related, to normative bases of decision-making. Yet in other situations, it is the physical environment that influences privacy concern and associated behavior, sometimes even unconsciously.

Some of the cues that influence perceptions of privacy are one's culture and the behavior of other people, either through the mechanism of descriptive norms (imitation) or via reciprocity. Observing other people reveal information increases the likelihood that one will reveal it oneself. Being provided with information that suggested that a majority of survey takers had admitted a certain questionable behavior increased participants' willingness to disclose their engagement in other, also sensitive, behaviors. Other studies have found that the tendency to reciprocate information disclosure is so ingrained that people will reveal more information even to a computer agent that provides information about itself.

Other people's behavior affects privacy concerns in other ways, too. Sharing personal information with others makes them "co-owners" of that information and, as such, responsible for its protection. Mismanagement of shared information by one or more co-owners causes "turbulence" of the privacy boundaries and, consequently, negative reactions, including anger or mistrust.

Likewise, privacy concerns are often a function of past experiences. When something in an environment changes, such as the introduction of a camera or other monitoring devices, privacy concern is likely to be activated. For instance, surveillance can produce discomfort and negatively affect worker productivity. However, privacy concern, like other motivations, is adaptive; people get used to levels of intrusion that do not change over time. In an experiment conducted in Helsinki, the installation of sensing and monitoring technology in households led family members initially to change their behavior, particularly in relation to conversations, nudity, and sex. And yet, if they accidentally performed an activity, such as walking naked into the kitchen in front of the sensors, it seemed to have the effect of "breaking the ice"; participants then showed less concern about repeating the behavior. More generally, participants became inured to the presence of the technology over time.

The context-dependence of privacy concern has major implications for the risks associated with modern information and communication technology. With online interactions, we no longer have a clear sense of the spatial boundaries of our listeners. Adding complexity to privacy decision-making, boundaries between public and private become even less defined in the online world where we become social media friends with our co-workers and post pictures to an indistinct flock of followers. With different social groups mixing on the Internet, separating online and offline identities and meeting our and others' expectations regarding privacy becomes more difficult and consequential.

● **Malleability and Influence**

Whereas individuals are often unaware of the diverse factors that determine their concern about privacy in a particular situation, entities whose prosperity depends on information revelation by others are much more sophisticated. With the emergence of the information age, growing institutional and economic interests have developed around disclosure of personal information, from online social networks to behavioral advertising. It is not surprising, therefore, that some entities have an interest in, and have developed expertise in, exploiting behavioral and psychological processes to promote disclosure. Such efforts play on the malleability of privacy preferences, a term we use to refer to the observation that various, sometimes subtle, factors can be used to activate or suppress privacy concerns, which in turn affect behavior.

Default settings are an important tool used by different entities to affect information disclosure. A large body of research has shown that default settings matter for decisions as important as organ donation and retirement saving. In addition to default settings, Web sites can also use design features that frustrate or even confuse users into disclosing personal information, a practice that has been referred to as "malicious interface design". Another obvious strategy that commercial entities can use to avoid raising privacy concerns is not to "ring alarm bells" when it comes to data collection.

Various so-called "antecedents" affect privacy concerns and can be used to influence privacy behavior. For instance, trust in the entity receiving one's personal data soothes concerns. Moreover, because some interventions that are intended to protect privacy can establish trust, concerns can be muted by the very interventions intended to protect privacy.

Control is another feature that can inculcate trust and produce paradoxical effects. Perhaps because of its lack of controversiality, control has been one of the capstones of the focus of both industry and policy-makers in attempts to balance privacy needs against the value of sharing. Control over personal information is often perceived as a critical feature of privacy protection. In principle, it does provide users with the means to manage access to their personal information. Research, however, shows that control can reduce privacy concern, which in turn can have unintended effects.

Similar to the normative perspective on control, increasing the transparency of firms' data practices would seem to be desirable. However, transparency mechanisms can be easily rendered ineffective. Research has highlighted not only that an overwhelming majority of Internet users do not read privacy policies, but also that few users would benefit from doing so.

Although uncertainty and context-dependence lead naturally to malleability and manipulation, not all malleability is necessarily sinister. Consider monitoring. Although monitoring can cause discomfort and reduce productivity, the feeling of being observed and accountable can induce people to engage in prosocial behaviors or (for better or for worse) adhere to social norms. Prosocial behavior can be heightened by monitoring cues as simple as three dots in a stylized face configuration. Whether elevating or suppressing privacy concerns is socially beneficial and critically depends, yet again, on context. For example, perceptions of anonymity can alternatively lead to dishonest or prosocial behavior. Illusory anonymity induced by darkness caused participants in an experiment to cheat in order to gain more money. In other circumstances, though, anonymity leads to prosocial behavior—for instance, higher willingness to share money in a dictator game, when coupled with priming of religiosity.

• Conclusions

Norms and behaviors regarding private and public realms greatly differ across cultures. Americans, for example, are reputed to be more open about sexual matters than are the Chinese, whereas the latter are more open about financial matters (such as income, cost of home, and possessions). And even within cultures, people differ substantially in how much they care about privacy and what information they treat as private.

If privacy behaviors are culture- and context-dependent, however, the dilemma of what to share and what to keep private is universal across societies and over human history. The task of navigating those boundaries, and the consequences of mismanaging them, have grown increasingly complex and fateful in the information age, to the point that our natural instincts seem not nearly adequate.

In this Review, we used three themes to organize and draw connections between the social and behavioral science literatures on privacy and behavior. We end the Review with a brief discussion of the reviewed literature's relevance to privacy policy.

Uncertainty and context-dependence imply that people cannot always be counted on to navigate the complex trade-offs involving privacy in a self-interested fashion. People are often unaware of the information they are sharing, unaware of how it can be used, and even in the rare situations when they have full knowledge of the consequences of sharing, uncertain about their own preferences. Malleability, in turn, implies that people are easily influenced in what and how much they disclose. Moreover, what they share can be used to influence their emotions, thoughts, and behaviors in many aspects of their lives, as individuals, consumers, and citizens. Although such influence is not always or necessarily malevolent or dangerous, relinquishing control over one's personal data and over one's privacy alters the balance of power between those holding the data and those who are the subjects of that data.

Insights from the social and behavioral empirical research on privacy reviewed here suggest that policy approaches that rely exclusively on informing or "empowering" the individual are un-likely to provide adequate protection against the risks posed by recent information technologies. Consider transparency and control, two principles conceived as necessary conditions for privacy protection. The research we highlighted shows that they may provide insufficient protections and even backfire when used apart from other principles of privacy protection.

The research reviewed here suggests that if the goal of policy is to adequately protect privacy, then we need policies that protect individuals with minimal requirement of informed and rational decision-making—policies that include a baseline framework of protection, such as the principles embedded in the so-called fair information practices. People need assistance and even protection to aid in navigating what is otherwise a very uneven playing field. As highlighted by our discussion, a goal of public policy should be to achieve a more even equity of power between individuals, consumers, and citizens on the one hand and, on the other, the data holders such as governments and corporations that currently have the upper hand.

(Excerpt from *Science*, January 30, 2015)

Exercises

I. Translate the following technical terms into English.

1. 行为科学 _____ 2. 实证研究 _____
3. 审查制度 _____ 4. 社会科学 _____
5. 隐私保护/侵犯 _____ 6. 态度调查 _____
7. 信息不对称 _____ 8. 有针对性的广告 _____
9. 基要主义者 _____ 10. 地理追踪系统 _____
11. 隐私悖论 _____ 12. 态度—行为二分法 _____
13. 政治倾向 _____ 14. 性取向 _____
15. 组合分析 _____ 16. 公共领域 _____

17. 现场/野外试验 _____   18. 社会资本 _____
19. 默认设置 _____   20. 亲社会行为 _____

II. Translate the following paragraphs into Chinese.

A first and most obvious source of privacy uncertainty arises from incomplete and asymmetric information. Advancements in information technology have made the collection and usage of personal data often invisible. As a result, individuals rarely have clear knowledge of what information other people, firms, and governments have about them or how that information is used and with what consequences. To the extent that people lack such information, or are aware of their ignorance, they are likely to be uncertain about how much information to share.

Two factors exacerbate the difficulty of ascertaining the potential consequences of privacy behavior. First, whereas some privacy harms are tangible, such as the financial costs associated with identity theft, many others, such as having strangers become aware of one's life history, are intangible. Second, privacy is rarely an unalloyed good; it typically involves trade-offs. For example, ensuring the privacy of a consumer's purchases may protect her from price discrimination but also deny her the potential benefits of targeted offers and advertisements.

Text C

What the "Right to Be Forgotten[1]" Means for Privacy in a Digital Age

By Abraham L. Newman

Despite the attention the right has received, we should not forget that it is just one innovative piece of a comprehensive privacy framework that must be implemented locally and enforced globally.

---

[1] The right to be forgotten is a concept discussed and put into practice in the European Union and Argentina since 2006. The issue has arisen from desires of individuals to "determine the development of their life in an autonomous way, without being perpetually or periodically stigmatized as a consequence of a specific action performed in the past".

Has Europe upended Internet privacy and free speech with its decision to create a "right to be forgotten"? In May 2014, the European Court of Justice ruled that European citizens have the right to request that search engines delink results to items that are considered inaccurate, irrelevant, or excessive. In the specific case, a Spanish citizen asked that Google remove a link to a newspaper account of his home foreclosure, a debt he had subsequently paid. In essence, this right acknowledges the stickiness of digital footprints. Photos, court records, and letters that used to get lost in file cabinets are neatly organized and accessible from our laptops. A childhood foible can haunt someone for a lifetime. The right reasserts our human instinct for redemption and forgiveness in the digital age.

This debate matters not only for how individuals use the right to be forgotten but also as a window into privacy protection more generally. In this essay, I review the controversies surrounding the European decision, describe where the right fits into a broader privacy protection framework, and discuss several implications of the debate. In particular, I argue that the right to be forgotten highlights differences in privacy protection across the globe, marks the emergence of distributed regulatory approaches, and underscores the importance of the international context for successful privacy policy.

Critics of the ruling claim that it will bring the demise of everything from Internet search to free speech. Search firms will be saddled with the excessive costs associated with processing requests to remove links, and individuals who exercise the right will disrupt free expression by altering search results. In the first five months since the ruling, Google has processed roughly 180,000 requests of which it accepted 40%. Although review requests are no doubt cumbersome for search firms, they do not appear to pose an insurmountable technical or financial burden. Google's stock in December 2014 is close to its 5-year high and enjoys a market capitalization at over $360 billion. Search firms from Google to DuckDuckGo must be prepared to respond to take down requests ranging from libel and defamation to copyright, which far outpace the right to be forgotten.

In terms of free speech, the European Court decision did not create a right that trumps all others but explicitly called for a balance between the right to be forgotten and other interests. Moreover, the effect of the decision on speech is limited as it does not require the deletion of the original content but rather the delinking of that content from search results. It takes us back to a world where people might have to go to a city hall or library to research past debts rather than instantly downloading them.

The stakes of data correction for consumers can be high. A 2012 study by the Federal Trade Commission in the United States estimates that one in four individuals have an error in their credit report that could affect their credit score. A 2014 lawsuit filed by the State of Mississippi against Experian, one of the largest credit reporting agencies in the United States, suggests that Experian produces reports with errors and that consumers have considerable difficulty correcting them. These errors affect the ability of millions of Americans to get competitive interest rates for home and auto loans, obtain security clearances, or pass rental applications.

When considering a privacy framework, however, rules about data correction and erasure are just one piece of the puzzle. Equally vital, if not more important, are rules that govern how data can be collected and then used by other parties that were not involved in the original data collection. Can a company like Uber (the app-based transportation network and taxi company) collect and store location data from individuals as they use their services; can those data be used for purposes other than securing transportation; can it then share that data with other companies or the government; and can it store data even after a customer has cancelled an account? The right to be forgotten is just one piece of a comprehensive data privacy framework that would include rules surrounding data collection and how data are then used, analyzed, shared, and secured.

- **Privacy Is Not Dead**

The right to be forgotten is a potent reminder that Europe has developed such a comprehensive approach and stands in sharp contrast to U.S. privacy policies. The European Court of Justice's decision builds upon a coherent and robust privacy framework in both the European Union and its member countries that includes rules concerning the collection, use, and storage of personal data in the public and private sector. These rules are overseen by independent national regulatory agencies known as data privacy authorities. In this system, big players like eBay or IBM work closely with regulators and implement internal data privacy policies in order to prevent data privacy scandals.

In contrast to the European system, the United States has a fragmented, patchwork approach to privacy regulation. With the passage of the Privacy Act in 1973, the United States focused privacy rules on data collection by the federal government along with a limited number of regulations covering a varied and idiosyncratic set of private-sector activities. In addition, there is no single regulator dedicated to overseeing the implementation and enforcement of disparate regulations.

The different approaches to privacy in Europe and the United States shape how governments and firms process and share personal information. In the United States, for example, nearly 100% of the population has a private-sector credit report, including "positive" information ranging from income to purchasing patterns. These types of data are routinely used to construct predictive scores such as the Consumer Profitability Score or the Individual Health Risk Score. In France, roughly 3% of the population has a credit report, which details "negative" information, such as defaults or missed payments, and as a result, there are far fewer predictive scores. European privacy rules are not a panacea for the immense challenges posed by digital technologies, but they offer a strikingly different set of ground rules from which to begin the debate.

- **Distributed Regulation**

The right to be forgotten is part of a trend in privacy protection toward distributed regulation, in which regulators rely on individuals and firms to monitor and implement regulations. Transparency, accountability, and class action remedies encourage consumer advocacy groups to organize and hold firms and governments accountable to the rules.

At the same time as consumer groups press for action, private firms increasingly carry out remedies. Companies, such as Google or Microsoft, have been deputized to evaluate and implement delinking requests. Whereas involving companies in the solutions has the benefit of distributing the task of enforcement, it raises the real risk of delegating sensitive issues like free speech regulation to corporations.

The right to be forgotten's emphasis on distributed regulation is similar to data breach laws that emerged from state-level experimentation in the United States. These laws require firms to notify customers when their personal data have been lost or stolen. California was among the first jurisdictions to adopt such rules in 2002, which have now spread to all U.S. states except Alabama, New Mexico, and South Dakota. Europe adopted similar rules in 2013 for telecommunications and Internet service providers and will pass more encompassing rules as part of the General Data Protection Regulation, which will be adopted in 2015. These laws have had a number of important impacts. Firms that encrypt their data are exempt from notification requirements, and so they have increased the use of encryptions. Finally, they put firms in the position of providing an important remedy by making credit report checks available to affected customers. These distributed regulatory policies do not eliminate more traditional forms of regulation, such as direct oversight or sanction, but expand the toolbox.

- **Privacy Goes Global**

Describing the European and U.S. approaches to privacy separately misses the important ways in which data protection is increasingly international. In today's digital world, information flows routinely cross borders, and such data flows are carried out by a handful of large technology and telecommunications firms. Citizens from Germany to Brazil must trust largely American companies like Google or Cisco to protect their privacy rights.

The right to be forgotten demonstrates the limits of national data privacy systems in a world of transnational data flows. In the wake of the European Court of Justice decision, once a delinking request has been approved, Google removes the link in the national domain name environment (for example, google.de, in Germany), while maintaining the link in other domains (such as the global google.com). Critics have argued that this negates the right as the offending information is still available on other domain name platforms. In November 2014, European privacy regulators made this concern official by recommending that companies respect such right-to-be-forgotten decisions globally.

The mismatch between regulatory jurisdiction and the transnational flow of information is a much more general phenomenon that plagues privacy protection. As revealed by the recent scandal involving the U.S. National Security Agency, personal data are often transmitted across networks that span countries. As companies employ more sophisticated surveillance techniques such as Super Cookies, these concerns are not limited to covert government activity.

To address these challenges, Europe has attempted to extend its jurisdiction beyond its borders. An important component of the European privacy system is that it limits data transfers to jurisdictions like the United States that do not have "adequate" protections in place. This has led many major information technology firms, such as Amazon and Google, to quarantine data about European citizens within European data centers.

The right to be forgotten has energized a debate concerning privacy on the Internet. At the same time, it needs to be kept in context of a broader privacy framework that considers how personal information is collected, shared, used, and secured. The European model offers one such approach. And that approach has been widely adopted outside of the United States in countries ranging from Canada to Argentina. National or regional privacy solutions, however, will face considerable difficulties if other countries, like the United States, maintain weak privacy systems that give safe haven to data brokers. Distributed regulatory tools, such as the right to be forgotten, will depend in large part on Europe's commitment to defending its system globally even if that means enforcing its rules across borders and periodically taking U.S. firms to task.

(Excerpt from *Science*, January 30, 2015)

Unit 12
Science and Society

∽ Exercises ∾

▌ Reading Comprehension

Directions: Answer the following questions based on the information from the text.

1. What is your opinion about "privacy protection" mentioned in the text?
2. How does the United States differ from Europe in terms of privacy rules?
3. Describe the way in which privacy protection goes global.
4. What is the significance of privacy protection system in a digital age?
5. Which privacy scandal is presented as an example in the text?

Appendix

常见数字、数字符号和数学式英语表达

科技英语中经常用到的数字、数字符号和数学式的英语表达列举如下：

一、数字表达

- 1/2   a half or one half
- 2/3   two thirds
- 1/4   a quarter or one quarter; a fourth or one fourth
- 3/4   three fourths or three quarters
- 1/10   a tenth or one tenth
- 1/100   a hundredth
- 1/1234   one over a thousand two hundred and thirty-four
- 119/1200   one hundred nineteen over one thousand two hundred
- $1\frac{1}{2}$ one and a half
- $3\frac{1}{2}$ three and a half
- $2\frac{5}{6}$ two and five over six or two and five sixths
- $5\frac{1}{8}$ five and one eighth
- $125\frac{3}{4}$ one hundred and twenty five and three fourths
- $64\frac{38}{483}$ sixty four and thirty-eight over four hundred eighty-three
- 0.1   0 (zero/nought) point one
- 0.01   0 point 0 one or zero point zero one or nought point nought one
- 0.001   0 point 00 one or (nought) point nought nought one or zero point zero zero one
- 0.12   (nought) point one two
- 0.056   zero point nought five six
- 3.14   three point one four
- 38.38   thirty eight point three eight
- 1,000,000,000   one billion

Appendix

常见数字、数字符号和数学式英语表达

- 1,000,000,000,000   one trillion
- 15%   fifteen percent
- 0.5%   decimal five percent
- 4‰   four per mill

二、数字符号表达

- +   plus or positive
- -   minus or negative
- ×   is multiplied by or times
- ÷   is divided by or over
- ±   plus or minus
- =   is equal to or equals
- ≡   is equivalent to or is identically equal to
- ≈   is approximately equal to
- ( )   parenthesis or round brackets
- [ ]   square (angular) brackets
- { }   braces
- 《 》   French quotes
- ' '   single quotation marks
- " "   double quotation marks
- >   is more than
- ≥   is more than or equal to
- <   is less than
- ≤   is less than or equal to
- ∝   varies as
- ∵   since or because
- ∴   hence or therefore
- ...   ellipsis
- ∠   angle
- //   parallel
- &   ampersand or and

- π   pi
- △   triangle
- ⊥   perpendicular to
- →   arrow
- ∪   union of
- ∩   intersection of
- ∫   the integral of
- Σ   (sigma) summation of; the sum of the terms indicated
- ∏   the product of the terms indicated
- $|x|$ the absolute value of x
- $\bar{x}$ the mean value of x
- R'   R prime
- R"   R double prime or R second prime
- R_1 R sub one
- #   number...

三、数学式表达

- 1+2=3   One and two are three.
- 2+4=6   Two plus four equals six.
- 5+0=5   Five and nought is equal to five.
- 46+74+122=242   46, 74 and 122 added are (or make) 242. The sum (or total) is 242.
- 9−5=4   Nine minus five equals (or is equal to) four.
- 25−7=18   Seven from twenty-five leaves eighteen.
- 25,654−8,175=17,479   8,175 (taken or subtracted) from 25,654 leaves 17,479. The difference (or the remainder) is 17,479.
- 2×0=0   Two multiplied by nought equals nought.
- 1×1=1   Once one is one.
- 2×2=4   Twice two is four.
- 3×6=18   Three times (multiplied by) six is eighteen.
- 9÷3=3   Nine divided by three makes (or is equal to) three.
- 30÷5=6   Five into thirty goes six times.

Appendix

常见数字、数字符号和数学式英语表达

- $83 \div 8 = 10\ldots3$ 8 into 83 goes 10 times, and 3 remainder.
- $(a+b-c\times d)\div e=f$ *a* plus *b* minus *c* multiplied by *d*, all divided by *e* equals *f*.
- $(8+6\frac{1}{2}-3.88\times 4)\div 2\frac{1}{2}$ eight plus six and a half minus three point eight eight multiplied by four, all divided by two and a half.
- $a \perp b$ *a* is perpendicular to *b*.
- $a//b$ *a* is parallel to *b*
- $a:b$ the ratio of *a* to *b*
- $a:b=c:d$ *a* is to *b* as c is to d; the ratio of *a* to *b* equals the ratio of *c* to *d*.
- x^2 *x* square; *x* squared; the square of *x*; the second power of *x*; *x* to the second power
- y^3 *y* cube; *y* cubed; the cube of *y*; the third power of *y*; y to the third power
- y^{-10} *y* to the minus tenth (power)
- $a^2=b$ the second power of *a* is *b*; *a* square is *b*; a squared is *b*; *a* to the second power is equal to *b*.
- $a^2+b^2=c^2$ *a* squared plus *b* squared equals c squared.
- $\frac{x^3}{a}=y^2$ *x* raised to the third power divided by *a* equals *y* squared.
- $f(x); F(x); \varphi(x)$ function *f* of *x*
- $y=f(x)$ *y* is a function of *x*
- $\int_a^b$ integral between limits *a* and *b*
- $\sqrt[3]{a}$ the cube root of *a*
- $\sqrt[5]{x^2}$ the fifth root of *x* squared
- $\sqrt{518}$ the square root of five hundred and eighteen
- $\sqrt[3]{930}$ the cubic root of nine hundred and thirty
- $f(x)=cx^2+ax+b, c\neq 0$ The function of *x* equals *c* times *x* squared plus *a* times *x* plus *b*, where *c* is not equal to zero.
- $5 \in N$ Five is a member of *N*.

Answer Keys

参考答案

Unit 1

Text A

II. Vocabulary

● **Section One**

1. A 2. B 3. C 4. D 5. A 6. B 7. C 8. A 9. D 10. C

● **Section Two**

1. radiation that affects a particular area after a nuclear explosion has taken place
2. alteration in the inherited nucleic acid sequence of the genotype of an organism
3. the capability of producing offspring; the frequency with which something occurs
4. being in a situation where something dangerous might affect you
5. severe overheating of the core of a nuclear reactor resulting in the core melting and radiation escaping
6. repeating a scientific experiment to obtain a consistent result
7. a minium or starting point used for comparison
8. relating to or conferring immunity to disease or infection
9. marking a position on a map using instruments to obtain accurate information
10. indicated a reading of gauges and instrument

● **Section Three**

1. E 2. G 3. J 4. A 5. K 6. B 7. L 8. C

9. I 10. D 11. N 12. H 13. O 14. M 15. F

Text B

I. Translate the following technical terms into English.

1. reaction chamber
2. laser beam
3. hydrogen isotope
4. self-sustaining fusion reaction
5. fiber laser
6. preamplifier
7. kinetic energy
8. spark plug
9. simulation code
10. implosion
11. thermonuclear weapon
12. fuel capsule
13. xenon flash lamp
14. shock ignition
15. high-powered laser pulse
16. ultraviolet light
17. Inertial Fusion Energy
18. Magnetic Confinement Fusion
19. strobe light
20. deuterium/tritium

II. Translate the following paragraphs into Chinese.

把PETAL和LMJ相结合的实验将模拟恒星以及其他天体内部（聚变）发生的条件。研究人员还将使用强大的激光冲击给质子加速——该方法可以产生治疗癌症的紧凑型加速器。但让激光核聚变研究人员兴奋的是来自PETAL的短促、急剧爆炸的前景，它可以作为核聚变反应点火的火花塞。

他们期望这样使用PETAL可以让IFE研究人员避免NIF曾经遇到的一些障碍。任何IFE项目的关键元素都是燃料芯块——一种大小与胡椒籽相仿的用来盛放冷冻氘和氚（氢的同位素，是核聚变的燃料）的塑料球体。这种被放在反应室中央的塑料芯块被来自激光脉冲的极高温汽化后，会产生一种向内破裂的力量，把燃料挤压到铅密度的100倍，并将其加热到1亿度（开尔文），这一温度足以为核聚变点火。

在NIF与LMJ的武器研究实验中，研究人员通过把燃料芯块包裹在一个金属壳中间接向内引爆芯块，这种金属壳可以通过激光加热，反过来再用X光炸裂芯块。这种方法具有一些优势，它可以消除激光束中的瑕疵，而且在内爆过程中X光比紫外线更具优势，但它却使目标变得复杂与昂贵，这不是科学家想要的产生能量的方式。NIF研究人员也曾努力让这一方式发挥作用，但能量在转化成X光的过程中丢失了，而且内爆进展得也不顺利。

Unit 2

Text A

II. Vocabulary

● **Section One**

    1. C 2. D 3. D 4. B 5. A 6. A 7. B 8. C 9. C 10. A

● **Section Two**

1. parts of an electronic system which perform a particular function
2. the degree of highness or lowness of a tone
3. obtain information from a larger amount or source of information
4. a large collection of written or spoken texts that is used for language research
5. means of doing work as provided by the utilization of physical or chemical resources
6. qualities or features that someone or something has
7. a single occurrence of a process
8. expresses a thought, opinion, or idea by using particular words
9. a spoken word, statement, or vocal sound
10. information about the result of an experiment, etc.

● **Section Three**

    1. G 2. E 3. H 4. I 5. M 6. D 7. J 8. L

    9. B 10. O 11. F 12. N 13. K 14. C 15. A

Text B

I. Translate the following technical terms into English.

1. nanomedicine	2. biodegrade
3. nanobot	4. biocompatible material
5. atomic scale	6. nanoengineering
7. medicinal payload	8. side effect
9. stomach acid	10. magnetic field
11. ultrasonic wave	12. stem cell
13. preparation	14. bio-mechanical stress

15. micro motor
16. electrostatic charge
17. autonomous steering
18. external guidance
19. chemical reaction
20. self-propelling device

II. Translate the following paragraphs into Chinese.

磁场和超声波方式最主要的缺陷就是两者都需要外部引导，这在实际应用中极为不便，而且磁场和超声波在体内的穿透深度有限。开发一种自主型"微型电动机"来输送治疗药物能解决上述难题。

此种微型电动机依靠化学反应来推进，但其毒性却是个问题。以氧化葡萄糖为例，它是一种血液中的糖分子，可产生能用作燃料的过氧化氢。但是研究人员已经很清楚，这种方式不能长期使用，因为过氧化氢会腐蚀活体组织，这样葡萄糖就不能产生足够的过氧化氢来推动微型电动机。使用其他自然产生的物质，比如（胃里使用的）胃酸或者（血液和组织里面大量存在的）水来做能源可能更有前途。

然而，如何实现这些自动推进式设备的精确导航可能是一个更大的障碍。我们不能因为这些纳米粒子能移动到任何地方，就以为它们能准确到达研究人员指定的位置。自主操纵现在还不是一个理想的选择，但是确保纳米药物在到达了正确的位置时再生效却不失为一个变通方案。

Unit 3

Text A

I. Reading Comprehension

- **Section One**

1. Research in cognitive science and psychology shows that testing, done right, can be an exceptionally effective way to learn. Taking tests, as well as engaging in well-designed activities before and after tests, can produce better recall of facts—and deeper and more complex understanding—than an education without exams.

2. In the handful of studies that have been conducted so far, scientists have found that calling up information from memory, as compared with simply restudying it, produces higher levels of activity in particular areas of the brain. These brain regions are associated with the so-called consolidation, or stabilization, of memories and with

the generation of cues that make memories readily accessible later on.

3. Across several studies, researchers have demonstrated that the more active these regions are during an initial learning session, the more successful is study participants' recall weeks or months later.

4. Hundreds of studies have demonstrated that retrieval practice is better at improving retention than just about any other method learners could use.

5. In an article published in 2010 University of Texas at Austin psychologist Andrew Butler demonstrated that retrieval practice promotes transfer better than the conventional approach of studying by rereading.

6. It reported that the metacognitive skills of students in classes that used exam wrappers increased more across the semester than those of students in courses that did not employ exam wrappers. In addition, an end-of-semester survey found that among students who were given exam wrappers, more than half cited specific changes they had made in their approach to learning and studying as a result of filling out the wrapper.

7. The grades earned by the 901 students in the course featuring daily quizzes were, on average, about half a letter grade higher than those earned by a comparison group of 935 of Pennebaker and Gosling's previous students, who had experienced a more traditionally designed course covering the same material. Students who took the daily quizzes in their psychology class also performed better in their other courses. The daily quizzes led to a 50 percent reduction in the achievement gap among students of different social classes.

8. If the state tests currently in use in the U.S. were themselves assessed on the difficulty and depth of the questions they ask, almost all of them would flunk. That is the conclusion reached by Kun Yuan and Vi-Nhuan Le, both then behavioral scientists at RAND Corporation, a nonprofit think tank.

II. Vocabulary

● Section One

1. C   2. D   3. B   4. B   5. B   6. C   7. B   8. D   9. A   10. C

● Section Two

1. the central point of something, typically a difficult or unpleasant situation

Answer Keys

2. formally accuse someone of something or make an assertion that
3. a system or planned way of doing things, especially one imposed from above
4. a mental concept regarded as corresponding to a thing perceived
5. a standard against which things may be compared or assessed
6. application of a skill learned in one situation to a different but similar situation
7. read things again and make notes in order to be prepared for the exam
8. serious thought or consideration about a particular subject
9. become better at noticing, thinking, or doing something
10. conditions or requirements in a legal document

● **Section Three**

1. F 2. I 3. G 4. J 5. D 6. M 7. L 8. K
9. B 10. N 11. A 12. C 13. O 14. H 15. E

Text B

I. Translate the following technical terms into English.

1. randomized controlled trial
2. gold standard
3. data analysis
4. metric
5. standardized test
6. eye gaze
7. multimodal learning analytics
8. methodology
9. academic credentials
10. exploratory activity
11. academic achievement
12. research design
13. longitudinal study
14. big data
15. galvanic skin response
16. merit pay
17. rigorous science
18. case study
19. education psychology
20. cognitive science

II. Translate the following paragraph into Chinese.

　　近来流行的"发现式"学习就是让学生自己去发现事实，而不是从老师那里直接接受。在布里克斯坦于2009年创办的一个名为FabLab@School的教育研讨网络上，他和同事正试图揭示问题的核心：学生到底需要多少教导。父母可能并不想看到他们的孩子在学习中遭遇挫折，但是布里克斯坦认为，"某些程度上的挫折和失败是富有成效的，是学生学习的有效途径"。他和他的同事试图通过研究发现学生是首先通过听老师讲，还是首先通过探索性活动学到的知识更多。他把先听老师讲称为"讲解与练习"，"你先听讲，然后练习"。学生被分成两组：一组从先听课开始，另一组从探索活动开始。研究人员在好几项研究中重复了这一实验，得到了比较一致的结果：先练习的学生其表现优于先听讲的学生25%。"我们的看法是，如果你没有首先对问题进行哪怕一点点的探索就听老师讲，你甚至都不知道老师的讲课是要回答什么问题，"布里克斯坦说。

Unit 4

Text A

II. Vocabulary

● Section One

1. C 2. D 3. B 4. C 5. A 6. A 7. B 8. D 9. B 10. A

● Section Two

1. streaks or stripes of different color in wood, marble, etc.
2. forming or causing to form crystals
3. particulate matter that is carried by water or wind and deposited on the surface of the seabed and may in time become consolidated into rock
4. radioactive substance or particle undergoes change to a different form by emitting radiation
5. changed from a liquid or fluid into a solid and became hard
6. a continuous extent of land or water, especially a stretch of river between two bends, or the part of a canal between locks
7. establish or ascertain the date of an object or event
8. something a little different from others of the same type
9. the relative amounts of a particular substance contained within a solution or mixture or in a particular volume of space
10. the chemical or physical decomposition of something

● Section Three

1. D 2. E 3. F 4. G 5. A 6. B 7. C 8. I

Answer Keys

9. J 10. M 11. L 12. N 13. O 14. H 15. K

Text B

I. Translate the following technical terms into English.

1. paleontologist
2. geologist
3. sedimentary basin
4. field crew
5. topography
6. sedimentary layer
7. remote-sensing specialist
8. Landsat 7 satellite
9. electromagnetic spectrum
10. spectral band
11. mineral deposit
12. radiation profile
13. spectral signature
14. pattern recognition
15. artificial neural network
16. computational model
17. high-resolution satellite imagery
18. fossil site
19. image-analysis technique
20. geospatial predictive model

II. Translate the following paragraphs into Chinese.

  对于每个化石沉积层，我们都要计算资源卫星提供的六个电磁频谱波段的组合值，所以要确定已知沉积层具有与众不同的光谱特征可不是件容易的事。我们的问题本质上还是一个多维度模式识别的问题，在这一点上，我们人类远远不如计算机擅长。所以我们采用了所谓的人工神经网络——一个能学习复杂模式的计算机模型。

  我们的人工神经网络显示，这个盆地的化石沉积层确实具有共同的光谱特征，它还能轻易地将砂石沉积层和其他类型的地被如湿地、沙丘等区分开来。但是这个模型也有局限性。神经网络本质上就是一个分析性的"黑匣子"，它能区分不同的模式，但不能显示区分不同模式的真实因子。所以，尽管我们的神经网络能够轻而易举地区分化石沉积层和湿地、沙丘，却不能告诉我们不同地被的光谱特征在资源卫星数据（有理由相信这些信息能帮助我们进行定向研究）的六个波段上是如何被区分的。人工神经网络方法完全是基于对单个像素的分析，这是它的另一个局限性。其问题是，单个卫星像素覆盖的225平方米的范围并不一定与化石所在位置的尺寸完全对应：有些地点比单个像素大，有些又比单个像素小。如此一来，人工神经网络对潜在化石沉积层或者其他类型地被物的地点和范围的预测并不总是与实际情况相吻合。

Unit 5

Text A

II. Vocabulary

• **Section One**

   1. C   2. B   3. A   4. D   5. A   6. B   7. A   8. D   9. A   10. C

• **Section Two**

1. becoming brown by genes churning out enzyme in the fruiting body of certain plants
2. a breed, stock, or variety of a plant developed by breeding
3. eliminate
4. official authorization for something to proceed or take place
5. controls, influences, or regulates an action or course of events
6. a taxonomic category that ranks below species, its members differing from others of the same species in minor but permanent or heritable characteristics
7. preparation of cells obtained in an artificial medium containing nutrients
8. the order in which amino-acid or nucleotide residues are arranged in a protein, DNA, etc.
9. the production of offspring by a sexual or asexual process
10. causing to interbreed with one of another species, breed, or variety

• **Section Three**

   1. E   2. F   3. G   4. H   5. A   6. K   7. B   8. M
   9. C   10. N   11. D   12. O   13. J   14. I   15. L

Text B

I. Translate the following technical terms into English.

1. hardwood trees
2. deciduous forests
3. straight-grained wood
4. rot-resistant
5. pressure-treated lumber
6. preservative
7. blight resistance
8. hybridized tree
9. backcross breeding
10. fungus-resistant
11. transgenic tree
12. surrogate mother

13. plant pathologist   14. exotic pathogen
15. tree embryo   16. amino acid
17. caustic substance   18. acid-degrading enzyme
19. symbiotic relation   20. pollen

II. Translate the following paragraphs into Chinese.

  对大多数人来说，美洲板栗已渐成一段遥远的记忆。然而，我们并非只能眼睁睁地看着这样的事情发生。积累了数十年的研究成果告诉我们，科学可以将美洲板栗树以及它曾给人类和野生动物提供的所有资源带回我们身边。在与板栗疫病进行了长达一个世纪的徒劳对抗后，我们现在找到了两种初见成效的方法。第一种是古老的园艺技术——杂交：首先用体型小得多，并且具有抗真菌性的中国板栗树与美洲板栗树杂交，再尽可能地将杂交种与其他美洲板栗树进行"回交"。研究人员希望能用这种方法培养出既带有抗病基因，又具有美洲特点的板栗树。但是，回交育种法不仅不够准确，而且需要栽种数以千计的树木，进行很多代的杂交和选择，才能找到用于重建的树苗。

  有鉴于此，我和同事都将精力集中在第二个方案上，即通过远比传统育种更加精准的方法改变板栗树的DNA，这样或许能更快地获得具有更强抗真菌作用的树种。我们从小麦、中国板栗树和其他植物中借取基因，插入美洲板栗的基因组中，培养出了数百棵转基因美洲板栗树。其中一些树木防御寄生隐丛赤壳菌的能力丝毫不逊色于它们的亚洲"亲戚"。如果这些树种通过美国农业部、美国环保署和美国食品及药品管理局的审批（这在五年之后就可能实现），它们就将成为第一批用于在原生环境中重建关键物种的转基因生物。

Unit 6

■ Text A

II. Vocabulary

● **Section One**

  1. B 2. C 3. A 4. D 5. A 6. A 7. D 8. B 9. A 10. A

● **Section Two**

  1. used to classify something in terms of its position on a scale between two extreme or

opposite points

2. the structure of the nervous system or brain perceived as determining a basic or innate pattern of behavior

3. take away something from something else so as to decrease the size, number, or amount

4. a projecting arm or handle that is moved to operate a mechanism

5. a feeling of sickness with an inclination to vomit

6. relating to muscular movement or the nerves activating it

7. easily influenced

8. being trained or accustomed to behave in a certain way or to accept certain circumstances

9. switch from one feature or state to another

10. a disease or abnormal condition

- **Section Three**

  1. E   2. F   3. G   4. J   5. A   6. B   7. C   8. O
  9. N   10. M   11. L   12. K   13. D   14. I   15. H

Text B

I. Translate the following technical terms into English.

1. phoneme
2. consonant
3. vowel
4. parentese
5. autism
6. sensitive period
7. intonation
8. mental processing
9. inflection
10. neural algorithm
11. attention deficit
12. social gating
13. neurotransmitter dopamine
14. distributional frequency
15. biomarker
16. dyslexia
17. learning mechanism
18. statistical pattern
19. vernacular
20. language acquisition

I. Translate the following paragraphs into Chinese.

我的实验室团队考查了父母语中能吸引婴儿的那些声音——音频较高、语速较慢、声调比较夸张的声音。我们发现在面临选择时，婴儿会选择听用父母语讲的短录音片段，而不是同一个母亲对其他成人说话的录音。高音调似乎更能吸引婴儿，抓住并保持婴儿的注意力。

父母语夸大了声音间的区别，使得一个音位与另一个音位彼此很容易被区分开。我们的研究表明，夸张的言语有助于婴儿将这些声音存储到记忆中。在近期的一项研究中，我的研究小组中的纳伊兰·拉米雷斯–埃斯帕萨（现在供职于康涅狄格大学）让婴儿在家里全天佩戴安装在轻质背心上的高保真微型磁带录音机。录音让我们进入了儿童的声音世界。研究表明，如果父母对婴儿用父母语说话，那么一年以后他们所掌握的单词数是父母未频繁使用婴儿语的婴儿所掌握的单词数量的两倍多。

Unit 7

Text A

II. Vocabulary

● **Section One**

1. B 2. D 3. C 4. A 5. D 6. A 7. B 8. A 9. D 10. A

● **Section Two**

1. thin coating or covering
2. unable to bend or be forced out of shape
3. produce a state of increased energy or activity in
4. quantity obtained by multiplication
5. a finishing coat applied to exclude moisture
6. the natural process of laying down a substance on rocks or soil
7. a datum about some physical state that is presented to a user by a meter or similar instrument
8. a soft heavy toxic malleable metallic element
9. anything serving to maintain separation by obstructing vision or access
10. leak from a container

● **Section Three**

1. J 2. K 3. O 4. M 5. L 6. H 7. N 8. C

9. F 10. G 11. D 12. I 13. B 14. A 15. E

Text B

I. Translate the following technical terms into English.

1. graphene
2. optical property
3. highly active catalyst
4. molybdenum disulfide (MoS2)
5. molecular beam epitaxy (MBE)
6. amorphous silicon
7. logic circuit
8. photon
9. electron mobility
10. hexagonal lattice
11. ultrathin semiconductor
12. two-dimensional material
13. flash memory device
14. electronic property
15. weak bond
16. condensed matter physicist
17. allotrope
18. photodetector
19. nanometer
20. phosphorus atom

II. Translate the following paragraphs into Chinese.

然而石墨烯让研究者们看到了电子学的一片新天地。他们看到了类似的材料具有新奇的光学性能和电学性能。并且，由于二维材料很薄，且大部分是透明的，它们在创造韧性好且透明度高的电子产品方面勾勒了一幅美好的前景。这种电子产品能够产生好莱坞多年前在电影里虚构的那种透视效果。从那以后，研究者们在这块领域里搜索查找，以期有更多的新发现。

通过寻找自然形成的二维薄片材料，材料科学家发现了能自然生成三维结构的原子层的稳定方式，并在此基础上制备出了几十种新的二维材料，更多的二维材料也在研发中。他们已经制备出单层硅（硅烯）、单层锗（锗烯）和单层锡（锡烯）。他们用氮化硼制作出一种绝缘体，它和石墨烯一样具有六边形网孔的晶格结构。他们还制备出单层的金属氧化物，这种金属氧化物可以用做控制某些化学反应的高效催化剂。他们甚至在这些薄片材料里束缚了水分子，尽管这样做目前有什么用处还不清楚。

Answer Keys

参考答案

Unit 8

Text A

II. Vocabulary

● **Section One**

1. B 2. B 3. D 4. D 5. B 6. B 7. C 8. D 9. A 10. A

● **Section Two**

1. an unmanned exploratory spacecraft designed to transmit information about its environment
2. a spacecraft designed to transport people and support human life in outer space
3. the first stage of a multistage rocket
4. an allowable amount of variation of a specified quantity, especially in the dimensions of a machine or part
5. seal consisting of a flat disk placed to prevent leakage
6. a tool with a pointed end or cutting edges for making holes in hard substance
7. the amount that could be achieved or produced in a specified time
8. facilities for a spacecraft to joining with a space station or another spacecraft in space
9. a substance used for sticking objects or materials together
10. the propulsive force of a jet or rocket engine

● **Section Three**

1. K 2. C 3. N 4. D 5. F 6. H 7. E 8. A
9. B 10. G 11. I 12. O 13. M 14. J 15. L

Text B

I. Translate the following technical terms into English.

1. compliant design
2. windshield wiper
3. engineering paradigm
4. seamless FlexFoi
5. robot snake
6. rigid structure
7. fiber-optic network
8. output displacement
9. hydrostat
10. shape adaptation/morphing

11. air load   12. give
13. electrostatic motor   14. hinge
15. weather testing   16. contracture
17. lumped compliance   18. distributed compliance
19. geared transmission   20. wind-tunnel test

II. Translate the following paragraphs into Chinese.

　　我开始研究柔性一体化设备并非因为它们看起来像有趣的新奇物，而是因为在某些应用中，不含组装的设计是必需的。以研究汽车传动装置这样的大型机械系统为起点，我开始了自己的研究工作。然而，在二十世纪九十年代早期，我发现自己研究的只是小型设备——微机电系统而已。这是那个时代的大环境。当时电信公司正开始研发光纤网络的光开关，他们利用微型电机迅速改变反射面的角度，使光信号朝一个方向或其他方向发送。不久以后，我开始反复研读沃格尔的论文并探索柔性设计。我与史蒂文·罗杰斯和他的团队在桑迪亚国家实验室微系统部门开始合作一个项目，在那里，我们研究出了一项效果似乎很完美的一体化设计。

　　桑迪亚需要研发一台有足够输出位移的线性马达——至少要达到十微米。然而，这种静电马达的制造限制使得它们的位移限制在两微米。我清楚自己不能简单地进行微型化处理，比如，对齿轮传动装置做简单的微型化处理。即使我们能找到一个有足够人手的团队来组装尺寸在一至两微米范围内的齿轮、铰链和轴承，这样组装出来的设备对于现代工程而言会显得凌乱而庞大。相对于微机电系统的规模而言，有着十分之一微米宽间隙的设备就像"万能工匠"一样有用。此外，微机电系统设备是成批生产的，这和集成电路的方式一样：在一块如拇指指甲大小般的面积上组装了数以万计的元件。综合考虑到这些情况，我设计了一台一体式的运动放大器，当它与静电马达结合时，能产生二十微米的输出位移。

Unit 9

Text A

II. Vocabulary

● Section One

Answer Keys

1. A 2. D 3. B 4. B 5. C 6. C 7. C 8. C 9. B 10. C

- **Section Two**

    1. a device that follows and records the movements of someone or something
    2. systems, processes, departments etc. that operate in isolation from others
    3. programs of software designed and written to fulfill particular purposes of the user
    4. relay over the Internet as a steady, continuous flow
    5. not physically existing as such but made by software to appear to do so
    6. a group of related hardware units or programs or both, especially when dedicated to a single application
    7. record something in written, photographic, or other forms
    8. a point in a network or diagram at which lines or pathways intersect or branch
    9. of or relating to, or denoting a visual image
    10. make something visible to the eye

- **Section Three**

    1. M 2. F 3. G 4. L 5. E 6. D 7. C 8. N
    9. J 10. B 11. A 12. H 13. O 14. I 15. K

Text B

I. Translate the following technical terms into English.

1. integrated circuit
2. central processing unit
3. memristor
4. transistor
5. tri-gate transistor
6. computational performance
7. Dennard scaling
8. outsource
9. cache memory
10. power-hungry
11. random-access memory
12. quantum-mechanical effect
13. logical operation
14. video footage
15. flash memory
16. memory bank
17. heterogeneous computing
18. memory hierarchy
19. neurosynaptic core
20. linear sequence

II. Translate the following paragraphs into Chinese.

根据惠普的说法，这三个东西——总称为分级存储器体系，其中，静态随机存

· 341 ·

取存储器在顶层，而硬盘位于底部——主要负责工程师在克服摩尔定律局限性的过程中所产生的大部分问题。如果没有高速、高容量的内存来存储字符并快速传送的话，再快的CPU也毫无用处。

为打破这种"记忆墙"，加州帕洛阿尔托的惠勒的团队一直在设计一种新型的电脑——The Machine，该机器形成一个统一的存储器层，这样就完全避开了分级存储器体系。内存层次结构分级存储器体系中的每一层都有其长处和弱点。静态随机存取存储器的速度极快（所以它可以跟上CPU的运行速度），但耗电且容量小。主内存，或动态随机存取存储器（DRAM），速度快、密度大且耐用——这是优点，因为它是计算机执行应用程序的工作平台。当然，电源切断时，主内存所存储的数据会全部消失，这就是需要"非易失性的"闪存和硬盘等存储介质来长期保存数据的原因。它们具有高容量和低功耗的优势，然而它们速度极其缓慢（而闪存的磨损速度相对较快）。因为每个存储介质中有重叠的权衡，现代计算机将它们连接在一起，因此CPU可以在分级存储器体系中有效传送数据。"这绝对是个工程奇迹，"惠勒说，"但这也是一种巨大的浪费。"

Unit 10

Text A

II. Vocabulary

- **Section One**

    1. B 2. C 3. A 4. A 5. C 6. A 7. D 8. B 9. D 10. A

- **Section Two**

    1. acts of listening to evidence in a court of law or before an official
    2. killing someone with a fusillade of bullets or other missiles
    3. the extent to which a person or organization that attracts public notice or comment
    4. the quality of being easily shaped or moulded
    5. action or intervention, especially such as to produce a particular effect
    6. formed the boundary of or enclosed
    7. test a scheme, project etc. before introducing it more widely
    8. make available for general viewing or purchase
    9. delve into an abundant source to extract something of value, especially information or

skill

10. write program instructions

● **Section Three**

1. E 2. H 3. I 4. G 5. K 6. C 7. D 8. B
9. A 10. F 11. N 12. J 13. O 14. L 15. M

Text B

I. Translate the following technical terms into English.

1. think tank   2. self-plagiarism
3. cold fusion   4. drone
5. chromosome   6. DNA base
7. medical jargon   8. computing power
9. text analytics   10. search protocol
11. query   12. peer review
13. basic equation   14. biomedical literature
15. similarity score   16. ethics review
17. reproducibility of results   18. dynamic electronic corpus
19. thesis proposal   20. patent due diligence

II. Translate the following paragraphs into Chinese.

我们最常用来搜寻的数据库是Medline，它的管理机构是隶属于美国国家卫生研究院的国家医学图书馆，收藏了医学领域里所有的生物学研究，包括来自数千种同行评议期刊的数百万篇研究论文的标题与摘要。在Medline上有一个可用关键字查询的搜寻引擎，因此只要输入几个关键字，例如乳癌基因，就会找到不少结果，而且常附有全文链接。但是我才刚转行研究生物医学，对于很多研究该从何下手毫无头绪。

eTBLAST的初期版本光是从Medline比对几百个单词的段落就得花数小时，但确实管用。我通过eTBLAST开始读懂科学论文，逐段掌握内容要点；我可以把某位研究生的开题报告丢进去，快速得知相关文献。我和研究伙伴甚至与谷歌公司谈过，要把软件卖给他们，可惜他们回复说这并不适合他们公司的商业模式。

后来事情有了奇怪的转变。好几次，我发现学生论文开题报告里的文字和其他未注明出处论文里的文字一样；这些学生因此接受了道德教育，而我则有了一个改变职业生涯的研究主题：有多少专业生物医学文献涉及抄袭？

Unit 11

Text A

II. Vocabulary

Section One
1. C 2. A 3. B 4. A 5. A 6. C 7. D 8. C 9. D 10. D

Section Two
1. the message or signals transmitted through a communication system
2. an amassed store of useful information or facts, retained for future use
3. a computer or computer program which manages access to a centralized resource or service in a network
4. establishments at which something is produced or processed
5. an intentional disclosure of secret information
6. quantities or supplies of something kept for use as needed
7. a distinctive patter, product, or characteristic by which someone or something can be identified
8. conducting a systemic review of
9. the conceptual structure and logical organization of a computer or computer-based system
10. fulfill an obligation

Section Three
1. G 2. K 3. A 4. E 5. I 6. O 7. J 8. L
9. M 10. N 11. F 12. D 13. C 14. B 15. H

Text B

I. Translate the following technical terms into English.

1. cyberattack
2. embedded pacemaker
3. denial-of-service
4. spam
5. botnet
6. cybernetics

Answer Keys

7. insulin pump
9. point-of-sale system
11. security socket layer
13. cyberespionage
15. cybersecurity
17. bug bounty
19. malware

8. uranium-enrichment centrifuge
10. cryptographic key
12. decrypt
14. icon
16. social media account
18. distributed immune system
20. two-factor authentication

II. Translate the following paragraphs into Chinese.

　　接下来的几年可能会很糟。越来越多的数据被泄露，可以肯定，我们会看到一个关于在数字领域中我们应该给政府让步多少控制权来换取信息安全的激烈争论。事实上，保护网络安全需要从许多领域寻找解决方案：技术、法律、经济和政治。它也取决于我们普通公众。作为消费者，我们应该要求公司确保他们的产品更加安全。作为公民，当政府刻意削弱安全时，我们应该让政府承担责任。当然，个人在面临潜在风险时，我们有责任保护自己的东西。

　　保护自己仅包括一些简单的步骤，如保持我们的软件更新，使用安全的Web浏览器，对我们的电子邮件和社交网络账号启用双因素身份验证。但也要意识到，我们每个人的设备是一个更大的系统中的一个节点，我们所做的小小的选择可以产生广泛的影响。再重复一遍，网络安全就像公共卫生：通过洗手和接种疫苗，就可以避免疾病的进一步传播。

Unit 12

Text A

II. Vocabulary

● Section One

1. D　2. B　3. A　4. A　5. B　6. B　7. B　8. D　9. C　10. C

● Section Two

1. people who are involved and influential in an area or activity
2. a point where two systems, subjects, or organizations meet and interact
3. a process in which environmental or genetic influences determine which types of

organism thrive better than others, regarded as a factor in evolution
4. actions or devices intended to disguise or misled
5. a piece of organized and concerted activity involving members of the armed forces of the police
6. chemical composition and properties of a substance
7. the quality of becoming more marked or intense
8. a small group of people brought together to discuss, investigate, or decide upon a particular matter
9. reduction of energy to a less readily convertible form
10. communities of animals of one kind living close together or forming a physically connected structure

● **Section Three**
1. K   2. F   3. O   4. G   5. J   6. B   7. M   8. L
9. N   10. E   11. A   12. H   13. D   14. I   15. C

Text B

I. Translate the following technical terms into English.

1. behavioral science
2. empirical research
3. censorship
4. social science
5. privacy protection/violation
6. attitudinal survey
7. information asymmetry
8. targeted advertisement
9. fundamentalist
10. geotracking system
11. privacy paradox
12. attitude-behavior dichotomy
13. political orientation
14. sexual orientation
15. conjoint analysis
16. public sphere
17. field experiment
18. social capital
19. default setting
20. prosocial behavior

II. Translate the following paragraphs into Chinese.

隐私不确定性的一个首要也是最明显的来源是信息的不完整性和不对称性。信息技术的发展使个人数据的收集和使用变得很隐蔽。其结果是，个人几乎不清楚自己哪些信息落入到他人、公司或政府手上，他们如何使用这些信息，并产生了怎样的后果。考虑到人们对这方面信息的缺乏程度或者对自身无知的了解程度,他们确实无法判断自己有多少信息可以分享。

两个因素加大了查明隐私行为潜在后果的难度。第一,一些隐私伤害是有形的,比如和身份信息窃取相关的金融损失,而其他很多则是无形的,比如个人生平为陌生人所掌握。第二,隐私保护并不总是好事,它总是和利益权衡交织在一起。比如,保护消费者的购物信息可能令他们免受价格歧视,但也剥夺了他们从有针对性的报价和广告投放中获益的权利。